W9-BRM-057

LAND TREATMENT OF MUNICIPAL WASTEWATER

Vegetation Selection and Management

Edited by

FRANK M. D'ITRI

230 Collingwood, P.O. Box 1425, Ann Arbor, Michigan 48106

Library of Congress Catalog Card Number 81-69071
ISBN 0-250-40508-3

Manufactured in the United States of America

Butterworths, Ltd., Borough Green, Sevenoaks,
Kent TN15 8PH, England

This book brings you practical and theoretical treatment of an extremely important and timely subject. The work of experts from academic and governmental backgrounds, it will prove a valuable contribution to anyone working with disposal problems, or involved with recycling or resource recovery. We are pleased to publish this edition for the use of the scientific and engineering community.

THE PUBLISHER

PREFACE

The 1977 Clean Water Act Amendments of P.L. 92-500 further strengthened the policy established in the original act of promoting the utilization of innovative and alternative waste management techniques, especially with regard to the municipal waste treatment program. One of the major alternative waste management techniques is land treatment of wastewater. Thus, more emphasis is presently being placed on this technology.

Land treatment of municipal wastewater used by slow rate or overland flow systems depends on the proper selection and management of vegetation for its success. Cultivars of major grain, forage, food, and fiber crops are bred specifically for different regions of the United States because of variations in growing season, moisture availability, soil type, incidence of plant diseases, and other factors. Therefore, a regional approach to the selection and management of vegetation for land treatment facilities is essential, and guidelines should be developed for each major climatic and crop producing region of the United States.

To develop guidelines for the north-central region, including the upper Mississippi and Ohio River drainage areas and the Great Lakes Basin, a conference/workshop was presented at Michigan State University February 23-25, 1981. As several major research projects in the region have been underway for a number of years, they had previously resulted in numerous papers and reports. Even so, few efforts had been made to compile data from these various projects into a set of guidelines for the regional selection and management of vegetation. Some of the major projects include the Water Quality Management Facility at Michigan State University, the Muskegon County Land Treatment Facility at Muskegon, Michigan; land application on crops by U.S. Department of Agriculture scientists in Minnesota; and forest management by U.S. Forest Service scientists in Michigan and elsewhere. Along with projects in adjacent regions, such as The Pennsylvania State University system and studies of overland flow by U.S. Army Corps of Engineers scientists in New Hampshire, they have resulted in the accumulation of sufficient data to formulate carefully researched and reasonably precise guidelines for vegetation selection and management.

The success of the conference was due to the interest and dedication of the many individuals and organizations that were involved, and it is my pleasure to acknowledge as many of them as possible.

First, I wish to thank the U.S. Environmental Protection Agency for sponsoring the conference and providing the necessary financial support. My special thanks are extended to Mr. Steven Polonscik, Great Lakes National Program Office, U.S. Environmental Protection Agency, Region V, Chicago, IL 60604, and Mr. Richard E. Thomas, Office of Water Programs Operation, U.S. Environmental Protection Agency, Washington, D.C. 20460, for their encouragement and support during the preliminary phases of planning as well as their participation in the conference itself.

Secondly, a grateful acknowledgment is extended to all of the authors whose research as well as general contributions of time, effort, and counsel made this volume possible.

Thirdly, I especially appreciate the support received within Michigan State University from the Institute of Water Research and Lifelong Education Programs. In preparation of this manuscript, I acknowledge and thank Terry Waters for her expert secretarial and typing skills and Lois Wolfson for her editorial assistance.

Frank M. D'Itri

Frank M. D'Itri is Professor of Water Chemistry, Institute of Water Research and Department of Fisheries and Wildlife, Michigan State University in East Lansing. With a PhD in analytical chemistry, his primary research emphasizes the analytical aspects of water and sediment chemistry, especially the transformation and translocation of phosphorus, nitrogen, heavy metals and hazardous organic chemicals in the environment.

Professor D'Itri is listed in *American Men of Science, Physical and Biological Science*. He is the author of *The Environmental Mercury Problem*, co-author of *Mercury Contamination: A Human Tragedy* and *An Assessment of Mercury In the Environment*, and editor of *Wastewater Renovation and Reuse*. In addition, Dr. D'Itri is the author of more than forty scientific articles on a variety of environmental topics.

Dr. D'Itri has served as chairperson for the National Research Council's panel reviewing the environmental effects of mercury and a number of symposia dealing with the analytical problems related to environmental pollution. He has been the recipient of fellowships from Socony-Mobil, the National Institutes of Health, the Rockefeller Foundation and the Japan Society for the Promotion of Science.

CONTRIBUTORS

*Brockway, Dale G.

Forest Ecologist, Gifford Pinchot National Forest, Forest Service, U.S. Department of Agriculture, 500 West 12th Street, Vancouver, WA 98660 (Former address: Water Quality Division, Michigan Department of Natural Resources, Stevens T. Mason Building, Box 30028, Lansing, Michigan 48909).

*Burton, Thomas M.

Associate Professor, Department of Zoology, Department of Fisheries and Wildlife, and Institute of Water Research, Michigan State University, East Lansing, Michigan 48824.

*Clapp, C. Edward

Research Chemist, U.S. Department of Agriculture, Agricultural Research; Professor, University of Minnesota, St. Paul, Minnesota 55109.

*Cooley, John H.

Principal Silviculturalist, North Central Forest Experiment Station, Forest Service, U.S. Department of Agriculture, 1407 South Harrison Road, East Lansing, Michigan 48823.

*D'Itri, Frank M.

Professor, Institute of Water Research and Department of Fisheries and Wildlife, Michigan State University, East Lansing, Michigan 48824.

*Dowdy, R. H.

Soil Scientist, U.S. Department of Agriculture, Agricultural Research; Professor, University of Minnesota, St. Paul, Minnesota 55109.

*Ellis, Boyd G.

Professor, Department of Crop and Soil Science, Michigan State University, East Lansing, Michigan 48824.

*Epstein, Lynn

Graduate Research Assistant, Department of Botany and Plant Pathology, Michigan State University, East Lansing, Michigan 48824.

*Erickson, A. Earl

Professor, Department of Crop and Soil Science, Michigan State University, East Lansing, Michigan 48824.

*Hook, James E.

Assistant Professor, Department of Agronomy, Coastal Plain Experiment Station, Tifton, Georgia 31794.

*Jacobs, Lee W.

Associate Professor, Department of Crop and Soil Science, Michigan State University, East Lansing, Michigan 48824.

*Jenkins, Thomas F.

Research Chemist, Earth Sciences Branch, Cold Regions Research and Engineering Laboratory, U.S. Army Corps of Engineers, Hanover, New Hampshire 03755.

*Kerr, Sonya N.

Environmental Research Analyst, School of Forest Resources, Pennsylvania State University, University Park, Pennsylvania 16802.

*Knezek, Bernard D.

Professor, Department of Crop and Soil Science, Michigan State University, East Lansing, Michigan 48824.

*Larson, William E.

Soil Scientist, U.S. Department of Agriculture, Agricultural Research; Professor, University of Minnesota, St. Paul, Minnesota 55109.

*Linden, Dennis R.

Soil Scientist, U.S. Department of Agriculture, Agricultural Research; Assistant Professor, University of Minnesota, St. Paul, Minnesota 55109.

*Martel, C. James

Research Environmental Engineer, Construction Engineering Research Branch, Cold Regions Research and Engineering Laboratory, U.S. Army Corps of Engineers, Hanover, New Hampshire 03755.

*Marten, Gordon C.

Research Agronomist, U.S. Department of Agriculture, Agricultural Research; Professor, University of Minnesota, St. Paul, Minnesota 55109.

*Myers, Earl A.

Professional Agricultural Engineer, Williams and Works, Inc., 164 West Hamilton Avenue, State College, Pennsylvania 16801.

*Palazzo, Antonio J.

Research Agronomist, Earth Sciences Branch, Cold Regions Research and Engineering Laboratory, U.S. Army Corps of Engineers, Hanover, New Hampshire 03755.

*Poloncsik, Stephen

Chief, Technology Section, Environmental Engineering Branch, Water Division, U.S. Environmental Protection Agency, Region V, 230 South Dearborn Street, Chicago, Illinois 60604.

*Safir, Gene R.

Associate Professor, Department of Botany and Plant Pathology, Michigan State University, East Lansing, Michigan 48824.

*Sopper, William E.

Professor, School of Forest Resources, Pennsylvania State University, University Park, Pennsylvania 16802.

*Sutherland, Jeffrey C.

Professional Geologist, Williams and Works, Inc., 611 Cascade West Parkway, S.E., Grand Rapids, Michigan 49506.

*Tesar, Milo B.

Professor, Department of Crop and Soil Science, Michigan State University, East Lansing, Michigan 48824.

*Thomas, Richard E.

Research Scientist, Municipal Construction Division, Office of Water Programs Operations, U.S. Environmental Protection Agency, Washington, D.C. 20460.

*Urie, Dean H.

Principal Hydrologist, North Central Forest Experiment Station, Forest Service, U.S. Department of Agriculture, 1407 South Harrison Road, East Lansing, Michigan 48823.

CONTENTS

Review of Crop Selection Research 1
R. E. Thomas

1. An Overview of the Current Status on the Selection and Management of Vegetation for Slow Rate Systems to Treat Municipal Wastewater in the North Central United States 5
D. G. Brockway, T. M. Burton, J. H. Cooley, F. M. D'Itri, R. H. Dowdy, B. G. Ellis, L. Epstein, E. A. Erickson, J. E. Hook, L. W. Jacobs, S. N. Kerr, B. D. Knezek, E. A. Myers, A. J. Palazzo, S. Poloncsik, G. R. Safir, W. E. Sopper, J. C. Sutherland, M. B. Tesar, R. E. Thomas and D. H. Urie

2. Selection of Irrigation System Design 19
J. C. Sutherland and E. A. Myers

3. Wastewater Crop Management Studies in Minnesota . . 35
R. H. Dowdy, C. E. Clapp, G. C. Marten, D. R. Linden and W. E. Larson

4. Crop Management Studies at the Muskegon County Michigan Land Treatment System 49
B. G. Ellis, A. E. Erickson, L. W. Jacobs and B. D. Knezek

5. Comparison of the Crop Management Strategies Developed from Studies at Pennsylvania State University, University of Minnesota, Michigan State University, and the Muskegon County Land Treatment System 65
J. E. Hook

6. Management Studies of Annual Grasses and Perennial Legumes and Grasses at the Michigan State University Water Quality Management Facility 79
M. B. Tesar, B. D. Knezek and J. E. Hook

7. Oldfield Management Studies on the Water Quality Management Facility at Michigan State University . . 107
T. M. Burton and J. E. Hook

8. Vegetation Selection and Management for Overland Flow Systems 135
A. J. Palazzo, T. F. Jenkins and C. J. Martel

9. Growing Trees on Effluent Irrigation Sites with Sand Soils in the Upper Midwest 155
J. H. Cooley

10. Tree Seedling Responses to Wastewater Irrigation on a Reforested Old Field in Southern Michigan . . . 165
D. G. Brockway

11. Studies of Land Application in Southern Michigan . . 181
T. M. Burton

12. Plant Diseases Associated with Municipal Wastewater Irrigation 195
L. Epstein and G. R. Safir

Author Index 205

Subject Index 211

ABBREVIATIONS AND CONVERSIONS

English to Metric - Metric to English

English Unit	Abbreviation	Multiplier English to Metric	Abbreviation	Metric Unit	Multiplier Metric to English
acre	acre	0.405	ha	hectare	2.471
acre-foot	acre-ft	1,233.5	m^3	cubic meter	8.11×10^{-4}
cents per thousand gallons	¢/1000 gal	0.264	¢/1000 l	cents per thousand liters	3.785
cubic foot	ft^3	28.32	l	liter	0.0353
		0.02832	m^3	cubic meter	35.311
cubic foot per second	ft^3/sec	28.32	l/sec	liters per second	0.0353
		0.02832	m^3/sec	cubic meter per sec	35.311
cubic inch	in^3	16.39	cm^3	cubic centimeter	0.0610
		0.0164	l	liter	61.025
cubic yard	yd^3	0.765	m^3	cubic meter	1.31
		764.6	l	liter	0.00131
cubic yards per acre	yd^3/acre	1.89	m^3/ha	cubic meters per hectare	0.529
degree Fahrenheit	deg F	0.555 (°F-32)	deg C	degree Celsius	1.8 C° +32
feet per second	fps	0.305	m/sec	meters per second	3.28
feet per year	ft/yr	0.305	m/yr	meters per year	3.28
foot (feet)	ft	0.305	m	meter(s)	3.28
gallon(s)	gal	3.785	l	liter(s)	0.264
gallons per acre per day	gad	9.353	l/day/ha	liters per day per hectare	0.107

gallons per capita per day	gcd	3.785	1/capita/day	liters per capita per day	0.264
gallons per day	gpd	4.381×10^{-5}	1/sec	liters per second	22,831
gallons per day per square foot	gpd/ft^2	1.698×10^{-5}	cu m/hr/sq m	cubic meters per hour per square meter	59,172
		0.283	cu m/min/ha	cubic meters per minute per hectare	3.534
gallons per minute	gpm	0.0631	1/sec	liters per second	15.85
gallons per minute per square foot	gpm/ft^2	2.445	$cu\ m/hr/m^2$	cubic meters per hour per square meter	0.409
		0.679	$1/sec/m^2$	liters per second per square meter	1.473
inch(es)	in	2.54	cm	centimeter	0.3937
inches per day	in/day	2.54	cm/day	centimeters per day	0.3937
inches per hour	in/hr	2.54	cm/hr	centimeters per hour	0.3937
inches per week	in/wk	2.54	cm/wk	centimeters per week	0.3937
million gallons	mil gal	3.785	Ml	megaliters (liters x 10^6)	0.264
		3,785.0	m^3	cubic meters	2.64×10^{-4}
million gallons per acre per day	mgad	0.039	$m^3/hr/m^2$	cubic meters per hour per square meter	25.64
million gallons per day	mgd	43.808	1/sec	liters per second	0.0228
		0.0438	m^3/sec	cubic meters per second	22.83
		3,785	m^3/day	cubic meters per day	2.64×10^{-4}

ABBREVIATIONS AND CONVERSIONS (continued)

English Unit	Abbreviation	Multiplier English to Metric	Abbreviation	Metric Unit	Multiplier Metric to English
mile	mi	1.609	km	kilometer	0.622
		1,609	m	meter	6.22×10^{-4}
parts per million	ppm	1.0	mg/l	milligrams per liter	1.00
pound(s)	lb	0.454	kg	kilogram	2.204
		453.6	gm	grams	2.204×10^{-3}
pounds per acre	lb/acre	1.121	kg/ha	kilograms per hectare	0.892
pounds per day per acre	lb/day/acre	1.121	kg/day/ha	kilograms per day per hectare	0.892
pounds per million gallons	lb/mil gal	0.120	mg/l	milligrams per liter	8.33
pounds per square inch	psi	0.0703	kg/cm^2	kilograms per square centimeter	14.23
square foot	ft^2	0.0929	m^2	square meter	10.76
square inch	in^2	6.452	cm^2	square centimeter	0.155
square mile	mi^2	2.590	km^2	square kilometer	0.386
square yard	yd^2	0.836	m^2	square meter	1.196
ton (short)	ton	907.2	kg	kilogram	1.102×10^{-3}
		0.907	metric ton	metric ton	1.102
tons per acre	tons/acre	2.24	metric tons/ha	metric tons per hectare	0.446
yard	yd	0.914	m	meter	1.094

REVIEW OF CROP SELECTION RESEARCH

Richard E. Thomas
Municipal Construction Division
Office of Water Program Operations
Environmental Protection Agency
Washington, D.C. 20460

INTRODUCTION

In 1972, the laws for water pollution control were amended to initiate a major federal role. This included a 3 year authorization of 18 billion dollars to fund water pollution control facilities which are publicly owned. Among the many provisions of the amended law was a requirement to encourage projects which would recycle wastewaters and nutrients in revenue producing enterprises involving silviculture.

In 1977, the federal law for water pollution control was again amended. Reclamation and reuse technologies were given more emphasis with a 10 year authorization of 45 billion dollars. This highlighted the amendments financial incentives for projects defined as innovative or alternative technologies. Among these, land treatment is emerging as a leading alternative. Interest has increased over the last decade and continues to generate a heavy demand for research to determine the environmental acceptability of a range of designs for land treatment projects involving silviculture and agriculture.

The U.S. Environmental Protection Agency's Office of Water Program Operations considers the state-of-the-art in the context of an operating construction program. It is not uncommon for the researcher, the regulator, and the user to diverge in their assessment of the same data base. In recognition of this, it is important for researchers and users to meet frequently for joint discussion of new inputs to the data base. Research conducted in the Great Lakes region in the last decade represents a significant new input and its discussion will strengthen the design and management of land treatment projects over a large geographic region. The presentation of research results and the subsequent assessment can influence the design and operation of both on-going and future land treatment projects.

THE PERSPECTIVE OF THE U.S. ENVIRONMENTAL PROTECTION AGENCY

Land treatment became a major consideration for wastewater management as a result of the federal legislation passed in 1972. This legislation identified land treatment as a recycling alternative which should be given serious attention in the national program to end the pollution of streams and lakes. The legislation mandated the evaluation of land treatment for all projects started after July 1, 1974. The mandate had been anticipated for several years, and an intensive research program was already well underway to assess the state of knowledge for the design and operation of these systems. The results led to the conclusion that dependable design options were available to implement land treatment as a recycling alternative. Program guidance was issued in July of 1974 and the U.S. Environmental Protection Agency embarked on an active program to use land treatment technologies to recycle wastewaters.

However, research would continue to improve our knowledge of design and additional program guidance would also be needed at frequent intervals. To consolidate the research findings and issue them as quickly as possible, the U.S. Environmental Protection Agency led the way to prepare a design manual on land treatment through the cooperation of an interagency group co-chaired by the U.S. Environmental Protection Agency and the U.S. Corps of Engineers. The manual, entitled PROCESS DESIGN MANUAL FOR LAND TREATMENT OF MUNICIPAL WASTEWATER, was issued in October, 1977, as a joint effort by the U.S. Environmental Protection Agency, the U.S. Corps of Engineers, and the U.S. Department of Agriculture. The best interpretation of current knowledge will be incorporated into the first revision which is scheduled for completion in October, 1981, and on-going research programs will continue to provide improvements for future revisions.

The statistics on the implementation of land application projects through the construction grants program indicate reasonable progress, but clearly more can be accomplished. Some 350 new projects were funded as a result of the 1972 act. The influence of the 1977 act has been significant as a result of the financial incentives. The number of projects involving land treatment has increased substantially. Projections based on the continuation of these financial incentives and increased understanding of land treatment indicate that over 1000 new systems are planned for implementation in the near future.

With the steadily increasing data based of research information that is being collected from the projects now underway or being planned or constructed, we are gaining a solid basis of understanding the process design and its reliability. These research gains support growing optimism that recycling alternatives to traditional wastewater treatment will become standard in the future. This philosophical position spearheads the U.S. Environmental Protection Agency's thrust to encourage use of land treatment as a wastewater management alternative that improves water quality, is cost competitive, saves energy, and recycles nutrients.

THE CURRENT STATUS ON THE SELECTION AND MANAGEMENT OF VEGETATION FOR SLOW RATE AND OVERLAND FLOW APPLICATION SYSTEMS TO TREAT MUNICIPAL WASTEWATER IN THE NORTH CENTRAL REGION OF THE UNITED STATES

by

D.G. Brockway, T.M. Burton, J.H. Cooley, F.M. D'Itri, R.H. Dowdy, B.G. Ellis, L. Epstein, A.E. Erickson, J.E. Hook, L.W. Jacobs, S.N. Kerr, B.D. Knezek, E.A. Myers, A.J. Palazzo, S. Poloncsik, G.R. Safir, W.E. Sopper, J.C. Sutherland, M.B. Tesar, R.E. Thomas, and D.H. Urie

INTRODUCTION

The 1977 Clean Water Amendments to Public Law 92-500 were enacted to strengthen the original policy of encouraging the utilization of innovative, alternative management techniques for the treatment and disposal of municipal wastewater. These alternative techniques include spray irrigation and overland flow land treatment systems which can be used individually or combined with lagoons. The lagoons serve as pretreatment systems for settling, microbial degradation of BOD, and/or for storage during periods of cold or wet weather. The proper selection and management of vegetation is critical for the efficient renovation of municipal wastewater by slow rate, overland flow, or spray irrigation systems.

A number of major research projects have been underway in the north central region of the United States for several years. Among them are: the Michigan State University Water Quality Management Facility (WQMF), the Muskegon County Land Treatment Facility at Muskegon, Michigan, land application on crops in Minnesota in a joint U.S. Department of Agriculture/ U.S. Army Corps of Engineers study, and land application on forests in the north central region by the U.S. Forest Service. When supplemented with findings from other projects in adjacent regions, such as The Pennsylvania State University system and the overland flow system at the U.S. Army Cold Regions Research and Engineering Laboratory (CRREL) in Hanover, New Hampshire, these projects have resulted in the accumulation of sufficient data to formulate guidelines to improve present wastewater renovation and vegetation management concepts. Nonetheless, until now very few attempts have been made to compile the results into a set of guidelines to select and manage vegetation.

SITE CHARACTERISTICS AND SYSTEM OPERATING PARAMETERS

The selection of land treatment relative to other methods depends on the size of the community and the expertise of local system operators, as well as the cost, quantity, and type of land available. The necessary expertise requires less formal training and emphasizes agronomic skills that are more apt to be available in small rural communities than the specialized skills needed to operate an advanced, centralized system. However, these land treatment techniques may not be economically feasible for large communities with limited access to reasonably priced rural land.

Once land treatment has been determined to be feasible, guidelines to operate and manage the system can be recommended based on local soil conditions and appropriate types of vegetation. Protecting the groundwater is a primary consideration. Over the past 18 years, land application has been studied on various research sites that span a broad range of soil types from spodosols to alfisols and mollisols with an array of cation-exchange capacities and organic matter.

To date, spray irrigation has been the most common land treatment method adopted in the north central United States. A wide variety of application techniques have been used, but center-pivot rigs or solid set distribution systems are most common. These systems can be used on flat, extensively undulating, and very steep land if the soil is moderately to highly permeable and the location has appropriate borders to protect the public from wind drift and/or spray aerosol. Flood irrigation, ridge and furrow, or other such practices may be more appropriate if borders are not available, or if the topography and soils are more suitable for these practices.

The responses of crops to slow rate wastewater application has been quite similar on all types of soil as long as the infiltration rate was not routinely exceeded. On sites with slow infiltration rates, for example fine textured soils that are only slowly permeable, an overland flow system may be more suitable depending on the topography. The land should be graded so that the water spreads uniformly, and the soil surface should be inspected frequently for incipient downslope channeling.

MANAGEMENT RECOMMENDATIONS FOR NITROGEN REMOVAL

Extensive research has been conducted on the relative quantities of phosphorus and nitrogen that can be removed by the soil and vegetation during various periods of the growing season. Whereas renovating the wastewater so as to preserve the quality of the groundwater is the primary goal, it sometimes conflicts with the secondary consideration of maximizing agricultural yields and profits. Frequently, recommendations for the land application of wastewater have been based on nitrogen mass balance data because nitrate-nitrogen can be readily leached from these systems at levels in excess of the drinking water standard. Therefore, the nitrogen application rates must be correlated with the levels of removal by the plant systems and soils to prevent excess nitrate-nitrogen from leaching into the groundwater.

While municipal wastewaters normally contain high levels of nitrogen, low nitrogen wastewaters are sometimes available through dilution (*i.e.*, from paper mills) or effluents from lagoon systems. Low nitrogen wastewaters, containing less than 10 mg/l of total nitrogen, can safely be applied to land to remove the phosphorus and solids and also to enhance crop production and groundwater recharge. High nitrogen wastewaters, those with more than 10 mg/l of total nitrogen, can be applied to the land if precautions are taken to assure nitrogen removal. Because of the close relationship to vegetation management, separate guidelines may be needed for low and high nitrogen wastewater systems. High nitrogen wastewaters were used to grow corn (*Zea mays* L.,), forage grasses, and legumes in The Pennsylvania State, Michigan State, Minnesota, and Muskegon projects. Because the Muskegon wastewaters were low in nitrogen, they were supplemented with commercial fertilizers to bring the levels to more than 10 mg/l to increase yields.

Application was usually restricted to the period from late spring to early fall, but the potential for winter irrigation was investigated at Michigan State University, Minnesota, and CRREL. During winter irrigation, the recharge volumes were greatest, and high nitrogen wastewaters were most likely to leach out of the root zone and into the groundwater aquifers. The greatest nitrogen losses coincided with periods of reduced plant growth or were prior to the development of significant root biomass for annual crops. Therefore, during the application of high nitrogen wastewater, actively growing vegetation with a well developed root biomass is required if nitrate-nitrogen leaching is to be reduced significantly. If irrigation is restricted to the growing season, nitrogen losses will be minimal because the leaching volumes are low during winter. Nevertheless, the limited amount of leachate which does occur may contain nitrate concentrations greater than 10 mg/l as nitrogen because of the absence of plant growth.

AGRONOMIC CROP MANAGEMENT SYSTEMS

Cultivars of major grain, forage, food, and fiber crops are bred specifically for different regions of the United States because of variations in growing seasons, moisture availability, soil type, incidence of plant diseases, and other factors. Therefore, a regional approach is essential to select and manage vegetation at land treatment facilities. Considerable data have been generated for slow rate systems in the north central region. However, only limited data are available on seeded crop yields and nutrient uptake at overland flow sites.

Because information about selecting and managing vegetation on overland flow slope systems is quite limited, research must be directed toward identifying, managing, and improving the functions that vegetation can perform. With proper study, this method of treating municipal wastewater should advance dramatically in the next few years. The overland flow process holds great promise because vegetation can perform multiple functions such as controlling erosion, delaying runoff (settling solids and promoting time-dependent phosphorus-soil reactions), entrapping solids, and removing nutrients by plant uptake. It can also serve as a denitrification medium, improve the aesthetics, and provide low cost treatment of wastewater, often with an economic return.

The vegetation management system must be selected to optimize wastewater renovation first and then the net economic return from various crops that might be grown on the site. Several cropping systems are adaptable, and their sale offers an opportunity to lower the overall cost.

Row Crops

Row crops such as corn are attractive because of their potentially high rate of economic return as grain or silage. However, the limited root biomass early in the season and limited period of rapid nutrient uptake can present problems. As a result, corn can remove extensive amounts of nitrogen only for a short period of time (four to six weeks) from approximately the fourth to the ninth week after the plants emerge. High nitrogen wastewater applied at other times often results in leachates containing nitrate levels greater than 10 mg/l as nitrogen. Prior to the fourth week, root biomass is too low to renovate the wastewater effectively, and after the ninth week plant uptake slows. However, during the rapid uptake period, corn efficiently removes nitrogen from percolating wastewater. In some studies, the corn yields were low unless other nitrogen fertilizers were applied during the critical uptake period. Conversely, if high nitrogen fertilizers were added, the possibility of nitrogen leaching into the groundwater was increased. If the wastewaters were low in nitrogen, however, corn was an excellent choice for wastewater renovation. For example, in the Muskegon studies, corn was grown year after year with no problem of excessive nitrate in the drainage water nor any interdrain leachate problems because the applied wastewater contained less than 10 mg/l nitrogen. Although not much nitrate was taken up by the corn, not much was leached either. Other row crops like soybean (*Glycine max* Nerr.) and sunflower (*Helianthus annuss* L.) are potential alternatives, but they have not been sufficiently studied to determine their relative value.

The management guidelines for the spray irrigation of high nitrogen (>10 mg/l) wastewater have to be very specific for corn. Either the application rate has to be adjusted to prevent leaching during low uptake periods, or there must be intercropping with annual or perennial forages. Intercropping may represent a suitable alternative to growing this grain alone. A dual system of rye (*Secale cereale* L.) intercropped with corn to maximize the period of nutrient uptake was tested in Minnesota and Michigan. For such dual corn-ryegrass cropping systems, rye can be successfully seeded in the standing corn in August, or after the harvest in late September. The growth of rye in the spring, before the corn is planted, allows early the application of high nitrogen wastewater. While planting the corn, a herbicide can be applied in strips to kill some rye so corn can be seeded in the killed rows. With the remaining rye absorbing nitrogen, less is leached during the early growth of the corn. Such intercropping allowed the successful use of corn as a high value crop. However, nitrate-nitrogen in excess of the drinking water standard was leached during the period that corn was being established in the partially killed rye if more effluent was applied per week than was lost through evapotranspiration. The corn-rye system is acceptable from a nitrogen renovation standpoint, and with careful attention to application rates, the technology is available for this option; but managing an intercropping system is more difficult than growing either crop alone. Perennial grasses and legumes can also be grown in a grass-corn intercrop system, but the practice requires careful control of the forage grasses and weeds. Where grasses are overkilled, the system behaves like a single corn crop; and the nitrogen renovation is ineffective. When too much grass remains, good nitrogen renovation occurs; but the corn yield is lowered.

Planting corn into the killed sods can also be beneficial because the sod serves as a winter cover to control erosion. The sod can be irrigated with wastewater in the spring before it is killed and again in the fall if the growth is sufficient. During the corn growing season, however, the problem remains that corn monoculture is inadequate to remove the nitrogen.

When both corn and hay are available for wastewater treatment, applications can be tailored to produce high corn yields by supplying water and a nitrogen supplement as needed. When the corn cannot use the water or when irrigation would increase nitrogen leaching, the wastewater can be diverted to the hay fields. Because such a system requires larger land areas, a primary consideration in the initial planning stage is land cost.

Non-Row Forage Crops

The tradeoff between good renovation and a cash return can best be attained by mixing corn and forage crops. However, by themselves, forage grasses and old field vegetation* can provide high levels of renovation with less operational and maintenance costs than corn. Grasses are suitable for overland flow as well as spray irrigation, and excessive leaching of nitrate-nitrogen is relatively easy to prevent.

Special care is necessary for overland flow systems. To prepare for a successful operation, the tract should be land-planed, and the soil must be finely graded to avoid channeling and to provide a firm seed bed. For rough grading, the equipment should follow the land's contour to minimize the potential for erosion; however, the final grading should be uniform for each section. The seeding equipment should also be operated on the contour to minimize the formation of downslope runnels, rills, and channels. Equipment with wide tires is strongly recommended to minimize rutting, both for

**The old field vegetation described here was a mixture of perennial grasses and weeds (mostly broad-leaved weedy perennial species) which volunteered in the area after the growth of crops of corn, clover (Trifolium spp.), grass, and alfalfa (Medicago sativa L.) were discontinued.*

planting and harvesting. For spray irrigation systems, typical agronomic practices for the region are usually satisfactory.

Before seeding, soil tests should be performed to determine the initial need for fertilization to establish the grass and to ensure its rapid growth. The rate of seeding for overland flow systems should be greater than for conventional hay fields and should follow the recommendations of the Soil Conservation Service for establishing sod in grassed waterways. For spray irrigation systems, a rate of seeding somewhere between typical agronomic practices and the procedure for grassed waterways should be suitable.

The seeding mixture for overland flow systems should include water-tolerant perennial grasses such as orchardgrass (*Dactylis glomerata* L.), reed canarygrass (*Phalaris arundinacea* L.), tall fescue (*Festuca arundinacea* Schreb.), or Kentucky bluegrass (*Poa pratensis* Leyss.), along with a "nurse" grass that will establish itself quickly and insure the stability of the surface of the soil while the other grasses become established. Perennial ryegrass has performed the nursing function well in many overland flow systems. Annual weeds also must be controlled to avoid competition with the beneficial grasses. For spray irrigation systems, the most economical species or mixture should be chosen from the four water tolerant species listed above. In this case, a "nurse" grass is not usually necessary.

After seeding the overland flow site, a mulch should be applied to hold the soil and maintain its moisture until the new grass is established and can be irrigated. Smaller sites can be sodded and the application of wastewater can begin immediately. Enough lead time should be allotted to establish the vegetation before irrigation begins unless conditions allow a light sprinkling to assist in establishing the plants. Also, less than normal loading rates should be applied initially to test the stability of the freshly planted slopes. Subsequently, the grasses should be inspected frequently and analyzed periodically to detect losses of vigor or other health problems. Mulching is unnecessary on spray irrigation sites with slopes less than 6 percent, but care should be taken to assure that a good stand is established before higher irrigation rates are initiated.

In spray irrigation systems, nitrogen is removed from the wastewater by the roots of the plants. Therefore, root depth is very important; and proper drainage is needed to assure an adequately aerated soil volume for deep rooting.

About 30 cm appears to be adequate for water tolerant grasses such as reed canarygrass. The less tolerant grasses, as well as alfalfa and corn, require deeper rooting if renovation is to be successful. The preferred maximum aerated depth of soil is about 40 to 60 cm for corn and 100 cm for alfalfa to allow adequate wastewater retention for nitrogen uptake in the root zone. To minimize leaching, the amount of wastewater can be reduced to lower the net downward movement out of the root zone. Good aeration also minimizes root disease in the crop.

Land application systems have given excellent removals of phosphorus and nitrogen. In the U.S. Army CRREL experiment at Hanover, New Hampshire, nitrogen and phosphorus removals were between 210 and 332 kg/ha and 27 and 48 kg/ha, respectively. These results demonstrate that, with careful management, significantly higher crop yields can be obtained on overland flow sites than on land used for normal forage agriculture. Similar yields and removal rates have been achieved on spray irrigation sites.

Another tradeoff that has to be considered is whether harvesting the forage crop enhances removal enough to justify the cost, as opposed to leaving the field unattended. Some studies indicate that excellent renovation of wastewaters can be achieved on unharvested fields of planted grasses or volunteer old field vegetation, at least for the first few years. Studies were not continued beyond 5 yr; however, it is likely that nitrogen renovation will decrease as soil nitrogen levels increase over time. Thus, harvesting to remove nutrients from the site is recommended in the long term.

Forage crops other than grasses can also be used for slow rate irrigation systems. For example, alfalfa is more valuable than grasses; but it does not tolerate excess water as well, particularly on fine textured soils. Root rot resistant varieties of alfalfa are acceptable to renovate wastewater, and they persist on artificially tiled or naturally well drained soils. However, increasing the rates of water application above the total amount lost through evapotranspiration and percolation may reduce the long term survival rate. Thus, re-establishment may be required more often, every 2 to 3 yr, rather than the 3 to 5 yr which is a standard agronomic practice in the region. As long as the alfalfa persists, however, the nitrate leaching levels should be acceptable. More nitrogen is leached from alfalfa than from grasses but less than from corn. Also, removing the cut forage as hay or greenchop assures that a major portion of the wastewater nitrogen is permanently removed from the land application system.

Harvesting schedules are important for forage crops. Maximum crop yields and nutrient removals result when forage grasses are harvested three or more times per year with the first harvest in late May or early June at the "early heading" or flower stage of spring growth. Alfalfa survives well on this system with the second harvest in mid-July or early August and a third harvest in September or early October. Old field vegetation also offers excellent renovation when it is cut and removed on this schedule. For all of the forage crops, the irrigated fields have to be dried for approximately one to two weeks before each harvest so the soil can support the harvesting equipment without damage from rutting and puddling. Therefore, the number of cuttings must be weighed against the cost of additional land to accommodate the down time during harvest. In addition, alfalfa has a lower nutrient uptake for two weeks following harvest, so the irrigation rate should not exceed the total amount lost through evapotranspiration and percolation. This consideration adds to the land required to irrigate this species.

Grasses or old fields offer the possibility of minimum management because they remove nitrogen from high nitrogen water even when left unharvested. In this case, the time of safe renovation is related to the period of biomass accumulation. After the growth of biomass peaks each year, nitrogen leaching increases. Cutting or mowing the fields, even without removing the cut biomass, can increase the period of active growth and effective renovation. However, because the nitrogen is not physically removed from the field as it is with hay, the long term nitrogen removal of such a mowing system is less certain.

Mowed old fields have adequately renovated nitrogen over many years. These and some of the overland flow systems have been dominated by quackgrass (*Agropyron repens* L.). Quackgrass can be an acceptable volunteer to replace the plantings of other grasses because it takes up large quantities of nitrogen and, if cut frequently, provides good quality feed for animals and forms a thick sod. Quackgrass, however, is a noxious weed and can only be established by sodding or natural selection because the sale of quackgrass seeds is illegal.

An alternative to the high cost to purchase land for even minimal management systems is to supply local farmers with wastewater for crop irrigation when privately owned farms are nearby. When the amounts of applied wastewater are near the evaporative losses from the crop, nitrogen leaching is reduced to a tolerable level as field crops of all kinds are usually able to tolerate more water than their evaporative losses.

Under these circumstances, either high or low nitrogen wastewaters can be safely applied to any row or forage crop. Thus, more wastewater may be supplied to farmers than the minimum required for their crops without reducing the yield. However, excessive rates of irrigation must be avoided as the extra water increases the leachates, especially when nitrogen fertilizer is being added.

RESEARCH NEEDS FOR SLOW RATE AND OVERLAND FLOW SYSTEMS

Overland Flow Systems

To supply more precise information for cost balancing, additional research on overland flow systems is needed to determine: (1) the optimum vegetation and soil management procedures to remove the most phosphorus and nitrogen, (2) the effects of various plant species on wastewater residence time in overland flow systems, (3) the bioaccumulation of toxic organics by plant uptake and the effects of these organics on the quality of forages removed for animal feed or maintaining viable crop-soil systems, (4) the water tolerance, yields and uptake of nutrients by various types of forages, and (5) the seasonal constraints on overland flow systems.

Slow Rate Systems

Corn

Corn has not reliably removed high concentrations of nitrogen in wastewaters. Yet, because of its economical desirability, research is needed to explain the following: (1) why corn roots are not able to take up nitrogen when the concentrations in the soil solutions are low; (2) why the inefficiency of corn in using nitrogen is affected by the methods of application, soil aeration, and root structure; and (3) what the maximum application rates are for corn before the yields are lowered and/or excessive nitrogen leaching occurs.

Other Crop Systems

Information is needed on all crop systems to determine: (1) the extent to which diseases lower the capacity of cropping systems to renovate wastewater; and (2) how diseases can be managed through variety selection, crop rotation, and irrigation frequency.

New Crop Systems

Because only a limited number of crop systems have been investigated for use on land treated with wastewater, additional research is needed to identify new plant species such as wheat, soybeans, dry beans, and others of economic importance to this region.

FORESTED SYSTEMS

Since the pioneering research began at The Pennsylvania State University in the early 1960's, interest has increased in utilizing forested sites for the renovation of wastewater. The early work indicated that coniferous and hardwood forest ecosystems could be effective living filters to improve the quality of wastewater by removing the nutrients. Generally, forest productivity was also increased by this process. Subsequent research in Pennsylvania, Michigan, New Hampshire, Washington, Georgia, and other states examined a wide variety of forest cover types and sites. Their diversity necessitates consideration of the vegetation, soil, hydrology, and climate unique to each. However, a similar pattern of response was common to most of them.

The greatest benefit was observed with tree seedlings. Irrigation increased their survival in abnormally dry years and, on some sites, made it possible to establish tree species that would not normally survive. As a group, *Populus* species and hybrids demonstrated the greatest growth response whereas pines demonstrated the least. Irrigation of established stands resulted in changes in soil nutrient and hydrological characteristics. Accelerated decomposition and elevated pH in litter, humus, and the A_1 soil horizon were nearly universal. The normal levels of BOD, suspended solids, phosphorus, pathogens, heavy metals, and trace refractory organic compounds in biologically treated effluent did not appear to be detrimental to growth. On deep, well-drained sandy soils, the only detrimental effects of wastewater irrigation have been boron toxicity in red pine and rapid weed growth in new plantings. Extensive ice breakage and blowdown occurred in one irrigated stand of pole-sized red pine growing on a heavy soil.

Slow rate irrigation is the only application technique that has been successfully used in forest ecosystems. A modification of overland flow was tried in fully mature mixed hardwoods; but extensive tree mortality occurred, and the wastewater phosphorus concentrations were not adequately reduced. Christmas trees have been irrigated on a trial basis with a center pivot irrigator, but solid set systems are more common. Winter irrigation of forest lands has been attempted on a limited scale in the northern United States. In addition to The Pennsylvania State University facility, where special nozzles were developed and are still being used effectively, winter irrigation has also been carried out at Greenville, Maine; Wolfboro, New Hampshire; West Dover, Vermont; Milton, Wisconsin; and Pack Forest, Washington, to name a few.

Wastewater irrigation rates and schedules should be managed so as to meet treatment criteria, with fertilization of vegetation to increase yields as a secondary objective. In forest ecosystems, the control of nitrate-nitrogen discharges to groundwater is usually the foremost design constraint. Forest stands of rapidly growing trees and understory flora most effectively remove nitrogen from wastewater. The intensively cultured hybrid and eastern cottonwood have been tried in the north central region, and they compare favorably with corn and forage grasses with regard to wastewater renovation. In addition, ten species of seedlings have been investigated on the Michigan State University WQMF with eastern cottonwood, white ash, and Scotch pine showing the most favorable growth responses.

Uptake in very young plantations of any species is low because they do not fully occupy the site. During this stage of development, vegetation other than trees serves as an important nutrient sink. Until the stand closes, a high level of management, approaching that for agronomic crops, is required to maintain the herbaceous ground cover without reducing tree survival or growth.

The optimum management for forest ecosystems irrigated with high nitrogen effluent utilizes rapidly growing tree species and hybrids that are intensively managed in short rotations. Changes in the hydrologic characteristics expected as a consequence of adding wastewater at each application site should be addressed in the design, operation, and maintenance procedures for the irrigation system. At some locations there may be no market for products of short rotation silviculture in small quantities, but a favorable cost-benefit ratio might be achieved by developing local uses for whole-tree wood chips such as for energy production and sludge composting. Growing Christmas and ornamental trees for community beautification is another management option. As previously mentioned, ground cover enhances the nutrient uptake, so intercropping is a possible management method to achieve additional cash return as well as to renovate high nitrogen effluents effectively.

Acquiring the necessary level of expertise for a program of intensive management is a major cost element. This may not be justified for small community systems unless this expertise can be acquired through local public service agencies or private service contracts and the part-time employment of individuals with the required skills and expertise.

To manage small volume wastewater treatment facilities (<0.5 MGD) utilizing forest land, it is preferable to minimize the operational costs and skill levels of the personnel by adopting an extensive rather than intensive vegetation management strategy. Typically, lagoon facilities with seasonal storage provide an effluent with minimum nitrogen concentrations during the growing season. An irrigation rate of 2.5 to 5.0 cm/wk, applied only during the growing season, would result in nitrogen loadings of 56 to 112 kg/ha/yr, depending upon the wastewater nitrogen concentration.

High nitrogen wastewater should not be applied to forests with little or no net accumulation of biomass, such as predominantly evenly-aged stands of mature trees. However, because the soil can remove phosphorus adequately, these stands are acceptable for low nitrogen-high phosphorus wastewater applications. The forest might be made acceptable for the application of high nitrogen wastewaters by modifications to introduce herbaceous plants and young trees. Red pine should not be irrigated with wastewater unless it occupies the only available site. If red pine is irrigated, it should be managed to accelerate conversion to more suitable species.

Reduced wind velocity in forests, visual screening by trees, and non-product values should be considered in selecting stands, specifying buffer zone dimensions, and controlling human access. Foresters' advice and assistance are helpful in formulating plans to develop wind and visual barriers and to optimize the distribution of species of appropriate sizes on and around the application site.

The forest ecosystem has little capacity to remove nitrogen when the plants are dormant. Therefore, if high nitrate wastewater is applied in winter, the nitrate concentrations in the leachate are likely to exceed those allowable under the drinking water standard. The difficulty and costs of winter irrigation vary with climatic conditions within the region. Therefore, the choice of year-round or warm-weather-only irrigation should be based on the requirements for nitrogen removal and a comparison of costs for wastewater storage versus operation and maintenance of the system during cold weather.

RESEARCH NEEDS FOR FORESTED SYSTEMS

Most research has focused on wastewater renovation in forest stands that are not managed or where standard silviculture procedures have been practiced to increase yields. The establishment and manipulation of stands to maximize nitrogen uptake has been only indirectly addressed. Except for the work at The Pennsylvania State University, studies of wastewater have been carried on for only 5 yr or less, a short time relative to the rate of change in natural ecosystems. A lack of long-term studies severely limits the opportunities to evaluate the effect of wastewater irrigation on species composition and sustained product yields.

Mismanagement of wastewater irrigation on forest land cannot be entirely avoided because the current knowledge of the process and its limitations is incomplete. Furthermore, facilities are sometimes not operated within appropriate design limits. The loss of the renovative capacity may not be recognized until major changes have taken place in the ecosystem. Then a collapse can result in the unacceptable discharge of pollutants. Some facilities have so little excess application area that the damaged system must be rehabilitated while it is still being utilized.

The kinds of research needed to solve these problems are listed below in three groups that indicate their relative urgency. First: (1) compare the amount and timing of nitrogen uptake by native herbaceous plant, brush, and tree species grown under existing forest stands; and (2) manage old field vegetation to maximize its nitrogen uptake while establishing tree plantations. Second: (1) develop operation and management procedures to rehabilitate ecosystems overloaded to the point of collapse; and (2) modify silviculture practices in existing stands to establish and maintain the mixture of herbaceous plants, brush, and tree species that will maximize nitrogen uptake. Third: (1) develop methods to harvest tree crops while maintaining nitrogen uptake; and (2) establish, manage, and harvest forage crops under natural stands and tree plantations.

CHAPTER 2

SELECTION OF IRRIGATION SYSTEM DESIGN

Jeffrey C. Sutherland, Williams & Works, Inc., 611 Cascade West Parkway, S.E., Grand Rapids, Michigan 49506
and
Earl A. Myers, Williams & Works, Inc., 164 West Hamilton Avenue, State College, Pennsylvania 16801

INTRODUCTION

The type of irrigation system to be selected depends on the quality of wastewater to be irrigated, restrictions on the quality of renovated wastewater leaving the site, and the characteristics of the site. Restrictions on irrigation effluent quality usually apply to one or more of the following parameters: nitrogen, phosphorus, suspended solids, biological oxygen demand, and coliform bacteria.

Land application history demonstrates that the use of the soil-vegetation complex has no specific pretreatment requirements (*i.e.*, treatment prior to land application) for domestic wastewaters. The increased levels of pretreatment required in many states and regions over the past twenty years or so are due to human choices related to public health, economics, or aesthetics. General concern over deteriorating environmental water quality and specific concern for the assurance of adequate management levels have encouraged most Great Lakes states to require pretreatment to the secondary level prior to land application.

The most cost-effective alternative for pretreatment for those municipalities having additional available land for four to seven months "winter" storage of wastewater is stabilization pond treatment. Stabilization pond effluents can be handled satisfactorily by most types of irrigation systems. The kinds of irrigation systems to be discussed herein are in the categories of slow rate infiltration and overland flow. Rapid infiltration and wetland application will be mentioned at times to highlight or contrast the slow rate and overland flow methods.

MAJOR SITE CONSIDERATIONS IN THE DESIGN OF IRRIGATION SYSTEMS

Introduction

The major site considerations affecting irrigation system choices are centered on soil, groundwater, and climate factors. Many types and sources of data required for the design of land application systems are listed in Table I. Land treatment considerations involve many principles that are familiar to small community and rural people. Thus, in continuity with their regular work, the county extension agent, the Soil Conservation Service (SCS) district conservationist, and the farmer can address their expertise to specific wastewater treatment needs. A frequent and challenging need is to identify the least area of acceptable land which will satisfy the treatment requirements and at affordable cost, with or without engineering improvements.

Soils

Texture is the most important soil characteristic due to its effect upon the hydrologic regime, soil renovation capability, crop production, and total land treatment system management. Medium to light-textured soils (loams, sandy loams, or loamy sands) can be preferable for land application systems, especially where cash crop production is to be emphasized, and perhaps under management conditions which are least dissimilar to conventional agricultural practice. Very heavy-textured soils, having little or no infiltration capacity, may be suitable for overland flow treatment, with substantial crop yields as a potential bonus. Well-drained soils that are too droughty for cropping without irrigation can be preferable either for slow rate or rapid infiltration in terms of overall economy of land purchase including unit cost and required land area.

For land application systems, soil survey reports prepared by the SCS in cooperation with agricultural experiment stations and local government units are the primary source of soil suitability information (McCormack, 1975). These reports list soil textures and thicknesses to a depth of about 150 cm, and usually include detailed interpretive information needed for land use planning. Soil series mapping is overlain on aerial photograph base maps, which facilitates relating soil restrictions to morphological features in the field.

Soils with high sand content usually have the higher infiltration and percolation capacities and are the best medium for applying any needed hydrological controls. Soils

Table I

Types and Sources of Data Required for Design of Land Application Systems

Types of Data	*Principal Sources of Data*
Wastewater	Local wastewater authorities
Soil type and permeability	SCS soil survey
Temperature (mean monthly and growing season)	SCS soil survey, NOAA, local airports, and newspapers
Precipitation (mean monthly, maximum monthly)	SCS soil survey, NOAA, local airports, and newspapers
Evapotranspiration and evaporation (mean monthly)	SCS soil survey, NOAA, local airports, newspapers, and agricultural extension service
Land use	SCS soil survey, aerial photos from the Agricultural Stabilization and Conservation Service and county assessors' plats
Zoning	Community planning agency, city and/or county zoning maps
Agricultural practices	SCS soil survey, agricultural extension services, and county agents
Surface and groundwater discharge requirements	State water agency and US EPA
Groundwater (depth and quality)	State water agency, USGS, and driller's logs of nearby wells

SCS = Soil Conservation Service
NOAA = National Oceanic and Atmospheric Administration
USGS = United States Geological Survey
USEPA = United States Environmental Protection Agency

high in clay, however, can provide better nutrient removal and wastewater renovation as effluent filters through the soil profile, provided the specific soils are sufficiently open to allow infiltration. A land tract must be evaluated in hydrologic and land treatment terms. Engineering improvement at the outset, with conscientious operation and maintenance practices throughout, are often needed to optimize a soil's natural attributes. Thick hay and grass crops improve both the hydrologic and renovation capabilities of the soils. Plant roots help improve drainage, soil structure, and tilth. Crop removal of nutrients from the soil chemical complex provides continuing renovation capacity. Good management provides for usable crops, cropping procedures, and winter cover crops which help maintain soil structure and permeability and help to reduce wintertime erosion.

Natural soils, even with low amounts of clay, are readily compacted under a range of moisture conditions, especially as imposed under frequent wastewater application. Planting and harvesting can cause extensive compaction, which reduces infiltration and percolation capability, and, according to Trouse (1977), significantly retards plant root growth. Attempts to relieve the symptoms of stress with more frequent irrigation only aggravate the problem. Thus, in the selection of an irrigation system for the land application of wastewater, the soils must be carefully evaluated from many points of view.

Groundwater

Groundwater considerations are necessary in the evaluation of potential wastewater application sites and the choice of irrigation method. Groundwater modification can affect crop growth and the site's renovating ability, as well as the possible uses of water leaving the site. Groundwater quality, depth, and direction of flow are usually considered when selecting an irrigation system. Present or potential potable aquifer recharge areas should be evaluated for slow rate land treatment, and the possible regulatory need to purge the ground (underdrains or wells) of irrigated wastewater before it leaves the treatment site should be fully explored at the outset because of the significant added cost factor involved. The highest water table should be no less than 92 cm beneath the land surface during peak irrigation application. High and fluctuating water tables, though not directly observable on the surface of the soil, can kill large sections of a plant's roots and cause much stress. A high water table can frequently be lowered and controlled with underdrains. Michigan requires pre-chlorination of irrigation wastewater which is to be managed with underdrains. Underdrains tend to reduce the bacterial and phosphorus removal capability of soils because they shorten the time and distance of filter flow travel before discharge into receiving waters. These impacts of underdrains can be critical in coarse-textured soil settings.

Climate

Climatic conditions of importance in irrigation system selection include temperature, evaporation, precipitation, and evapotranspiration. Temperature is important because it determines the length of the growing season and affects the amount of yearly evaporation. Freezing weather usually negatively affects the ease of operation, crop production, and the degree of wastewater treatment. Freezing weather inhibits or negates year-round operation in most midwestern situations. The longer the winter layover, the greater the irrigation land requirement for applying the wastewater within a shorter time period, and the greater the pond storage volume requirements. In humid areas, evaporation and precipitation are nearly equal and, therefore, have little effect on pond storage requirements. The soil erosion and surface runoff effects of precipitation, however, must be considered in system planning.

In arid areas, evaporation in excess of precipitation reduces pond storage requirements, but this factor also reduces the area that can be effectively irrigated for cash crops. Evapotranspiration in arid areas may be so high that crop damage occurs due to concentration of salts.

DESCRIPTION OF LAND APPLICATION SYSTEMS

General

Rapid infiltration, overland flow, and slow rate processes each need to be considered when choosing a specific land treatment type for any municipality. These three general processes differ extensively in their annual application amounts, averaging perhaps 46, 6, and 2.5 m, respectively. Besides considering pure economics, each system type needs to be viewed as to how it will fit into the community's needs and outlook.

Because this conference relates primarily to management of vegetation, only the slow rate and overland flow process systems will be discussed. These two systems will be discussed independently, with the slow rate presentation first. Emphasis will be given those factors which should be considered in choosing the type of system because much greater emphasis on selection and management of vegetation will be presented by other authors.

Slow Rate Process Systems

The slow rate wastewater land application process, usually referred to merely as irrigation, is the controlled release of wastewater onto land having sufficient infiltering capacity to take in the applied wastewater. The distribution of wastewater may be by surface spreading or by spraying. In the Great Lakes region, the annual application depth may vary from 0.9 to 2.5 m applied at levels from less than 2.5 to greater than 10 cm/wk throughout the growing season. Renovation of the wastewater usually takes place almost entirely in the root zone (plant uptake) or the top 0.8 to 1.6 m of soil through filtration, adsorption, and chelation.

The primary disadvantages of the slow rate system are the land area required and the high cost per hectare of installation. Advantages of the slow rate process include ready adaptability to traditional agricultural goals, including cash crop production; and performance levels that are least sensitive to operational changes so that treatment reliability under variable conditions is optimal (Pair, 1969). Many municipalities have chosen the slow rate process because of those advantages and the desire to use the maximum amount of wastewater and nutrients for growing cash-value crops. Crop removal often accounts for most of the nitrogen and about half of the phosphorus in wastewaters (Kardos *et al.*, 1974). Corn for silage or for grain and non-legume grasses are crops frequently grown under the slow rate process.

Surface Application Systems

Surface application includes various level-field and graded-slope approaches. It is often preferable to avoid graded surface irrigation methods such as corrugations and graded furrows, unless the wastewater facility is in an area where these kinds of field surfaces are commonly prepared. As a rule, the specialized equipment and expertise must be locally available for these systems to be operated efficiently and economically.

Flood irrigation of leveled areas for crop production can be used somewhat more easily and widely than the graded methods. Two items of paramount concern, however, are the possible reduction of infiltration and percolation rates when the soil is leveled, and the necessity for developing a method of applying the wastewater to fit the crop requirements and the hydrologic conditions of the finalized land profile. Extreme care must be maintained to prevent undue compaction during the construction process. Permeability rates of only about 10 percent of those listed by the SCS for undisturbed conditions should be used in the design, particularly if cutting to below agricultural soil depth is to be done in rough grading.

Spray Systems

Travelers. The traveling gun irrigation system consists of a large, single sprinkler, which distributes water over a rectangular area as it automatically moves across a field. The most frequently irrigated dimensions are approximately 100 by 400 m or 4 hectares. A 9.2 or 13 cm diameter drag hose supplies the water to the moving sprinkler unit; its length is usually half the length of the sprinkler travel distance, with the water source in the center of the field. The sprinkler unit may be self-propelled by pumped water or by a small internal combustion engine, or it may be moved by an auxiliary engine on a cable winch stationed at the end of the field. The auxiliary engine can also be used to wind the hose on a large reel for easy transporting to the next irrigation location.

The application rate in liters per minute depends on the size of sprinkler and the operating pressure, not on the speed of travel. Variation in travel speeds is used to apply different depths of water. A 1325 l/min gun moved at about 60 cm/min, when used with a 100 m lane spacing, applies about 2 cm of wastewater on 4 hectares in 10 hours. Thus, a traveler unit in a six day week can easily irrigate 25 hectares, which makes it a fairly low-cost per hectare system. Due to high labor requirements, however, traveler systems generally are more adapted to farmer-owned sites, where different fields may be irrigated each year, than to city-owned and managed one-site operations.

Center Pivots. Center pivot sprinkler distribution systems are continuously propelled in a circle around a fixed pivot by pneumatic, mechanical, hydraulic, or electrical power. The irrigation lateral pipeline is supported high enough above the ground to clear growing crops and is carried by mobile units (or towers) that are individually powered at speeds regulated to maintain the lateral in a straight line. Mobile units range from about 27 to 46 m apart; and by varying the number used, the area irrigated by one pivot system can range between 0.8 and 80 hectares.

Center pivots for distribution of effluent need to be carefully designed and managed for specific site conditions. This may mean using several units instead of one larger unit, higher pressure at the pivot supply end, larger pivot lateral pipe diameter, elimination of end gun use, or faster speed of travel reducing the amount applied each rotation. Good management, or lack of it, can make or break any irrigation system; however, this is more true for center pivot systems than most types. This is because the entire center pivot functions as a single unit even though it may operate over a wide range of soil and site conditions.

Due to their low cost per hectare and their relatively low labor requirements, center pivots frequently are chosen when site conditions are appropriate. These conditions include no woods, slopes from nearly flat to about 10 percent, soils of nearly uniform texture over the entire area, and a field shape that is nearly square.

Solid-Sets. A solid-set sprinkler distribution system is composed of a network of main lines, submains, and laterals. Sufficient laterals are required to spray the entire site without moving any lateral or adding any additional pipe. Control of water to each area is merely by a valve at the beginning of each lateral or control point. Solid-set systems may be portable, where all pipes are picked up and removed from the area at the end of an irrigation season; or permanent, where the pipes remain in place from year to year. Solid-sets are favored for odd-shaped areas, for fields with much soil variation, for wooded areas, and for sites with geologic constraints such as sink holes or steep slopes. A portable solid-set system would be appropriate where an area is irrigated one year and a different area the following year. Permanent arrangements usually are used for year-round irrigation and for forested areas, especially when freezing conditions are encountered.

Risers of 2.5 cm diameter frequently are used to carry water from the lateral pipe to the sprinkler heads. Riser lengths should be such that sprinklers are a minimum of 0.6 m above ground in wooded areas and 0.9 m in cropped areas. Wastewater sprinkler heads often have only one nozzle. This permits the maximum diameter hole for the required discharge and reduces plugging from solids in the effluent. Small holes plug too easily and large holes require too much pressure; therefore, orifices in the range of 0.48 to 0.80 cm usually are preferred for solid-set land application systems.

Slow Rate System Example

The five basic parameters used in most slow rate wastewater irrigation system designs are listed in Figure 1. Here these parameters and their subsequent relationships are shown as they pertain to the design of the buried solid-set system presently in use at Middleville, Michigan. Brief comments on each of these parameters follow.

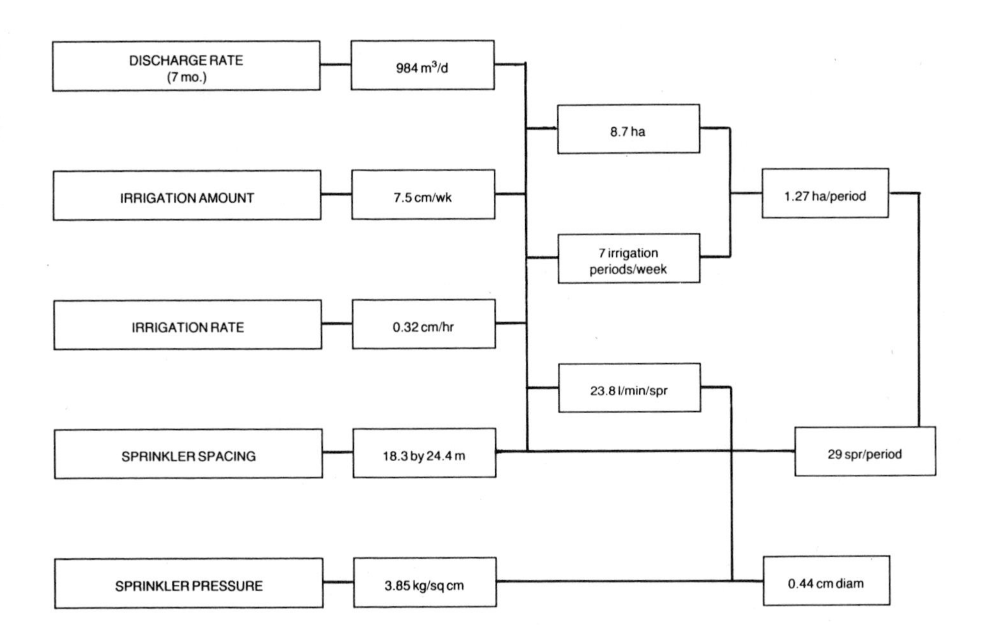

Figure 1. Summary of design parameters for Middleville, Michigan. Buried solid-set sprinkler land application system.

Communities usually produce wastewater continuously throughout the year; however, this flow varies in volume per day from season to season and in rate from one part of the day to another. This variation in flow rate, coupled with the general inability of land in the midwest to accept wastewater at a constant rate year-round, requires storage reservoirs or equalizing basins in most systems. The reasons for variation in the ability of land to accept continuous flows of wastewater include excessive precipitation periods, crop planting and harvesting operations, exceptionally cold periods, and periods during which crops cannot remove specific pollutants. Due to the uniqueness of wastewater, site, and climatic factors, the storage requirements are established on a site-by-site basis. At Middleville, five months of lagoon storage are being used.

The irrigation amount or weekly loading depth depends upon the available land area and the renovation and hydrologic capabilities of the site relative to the quality of the wastewater applied. Loading depth is usually expressed in centimeters of liquid distributed each week over the entire treatment area (Malhotra and Myers, 1975). This depth may vary from as little as 0.64 to over 15.0 cm/wk. The low loading would relate to a high strength wastewater that is irrigated on a soil of high clay content during cold weather or under sparse cover crop conditions. The high loading may be needed to supply sufficient nitrogen to corn on a sandy soil in mid-August when irrigating with low strength effluent. A 7.5 cm/wk irrigation amount was chosen for Middleville.

The design hourly application rate for typical agricultural irrigation is established for dry soil conditions. In liquid waste treatment and disposal, however, one often must irrigate during or after heavy rains, when crops are very young, in early spring or in late fall, and when the soil is near or at field capacity. To permit infiltration under these adverse conditions and to provide time for plant uptake and soil fixation of pollutants, it usually is preferable to irrigate in the 0.32 to 0.84 cm/hr range. An application rate of 0.32 cm/hr is being used at Middleville.

The spacing between sprinklers for solid-set systems usually ranges between 12.2 to 24.4 and 24.4 to 36.6 m (Myers, 1973). The first number refers to the spacing between sprinklers along the laterals, while the second refers to the distance between lateral lines. In open agricultural areas, a 24.4 by 30.5 m spacing frequently is used. The wider space between lateral lines is preferred for farming operations, which usually means lower capital and labor costs because fewer lines are required. Closer spacings of 18.3 by 24.4 m generally are used in forested areas to obtain reasonably uniform distribution of effluent. A spacing of 18.3 by 24.2 m was also chosen for Middleville because it fit well in the available open area, and it was the preferred spacing for erosion control.

The operating pressure at the sprinkler nozzle and the diameter of the nozzle are chosen so that the wastewater can be distributed uniformly. For a specific nozzle diameter, if the pressure is too high, the water will mist and drift extensively. If the pressure is too low, the discharge stream will not adequately be broken and an improper application pattern will result. Good distribution at Middleville is being attained with 3.85 kg/sq cm sprinkler operating pressure.

Surface Methods

Introduction

Overland flow and wetland flow are primarily surface water technologies. Overland flow is an "upland" approach wherein wastewater flows down evenly-graded "impermeable" grassed slopes on a schedule which alternates with frequent and regular rest periods. In wetland treatment, land slopes are nil or negligible and the surface may be highly uneven. Constant wetness is an acceptable feature, which allows considerable flexibility in irrigation schedules.

In their usual modes, these technologies may rely little on contact with soils for filtration and adsorption of impurities. Instead, soils are a substrate for microorganisms which attack solid and chemical impurities. Aerobic-anaerobic conditions ideally operate in both technologies to provide BOD removal, ammonia-nitrogen removal, and nitrification-denitrification. Vegetation acts as a physical filter for suspended solids and takes up nutrients.

In midwestern and New England states, cohesive glacial soils are frequently unsuitable for more conventional slow rate infiltration by means of spray and flood irrigation. In such natural settings, overland flow offers treatment advantages including conversion of raw (comminuted or settled) wastewater to better than secondary level during warmer months; and the capability of upgrading the quality of stabilization pond effluent before discharge to receiving waters. Natural wetlands, used selectively, will provide significant upgrading of pretreated wastewater while enhancing or insignificantly compromising natural wildlife and other values.

A weakness of both methods lies in phosphorus removal, which tends to be inefficient or impermanent because of the absence of substantial contact of wastewater with soils. In Michigan, the phosphorus disadvantage must be overcome for compliance with stream and lake water quality standards, and it has been overcome successfully in wetland-tertiary treatment for the Houghton Lake and Vermontville communities. At Houghton Lake, Michigan, reactive sapric-peat soils provide efficient removal of phosphorus. At Vermontville, Michigan, seepage through mineral soils, rather than surface flow-through, provides phosphorus removal. The present paper will deal no further with the wetland subject. For further information on the wetland methodology, see USEPA (1980a), Williams and Sutherland (1980), and Kadlec *et al.* (1979-81).

Overland Flow

Overland flow is comparatively little used at present. The total number of overland flow systems operating to treat municipal wastewater is probably less than an average of one per state. Initial confidence in the overland flow technology for wastewater treatment was developed largely through studies in the southeastern and southern United States (Peters *et al.*, 1980; Pollock, 1980; USEPA, 1977; USEPA, 1981). The most widely published studies with direct application to the Midwest are those by the U.S. Army Cold Regions

Research and Engineering Laboratory in Hanover, New Hampshire (Jenkins *et al.*, 1978; Martel *et al.*, 1980). Food processing wastewater treatment by overland flow, beginning in the early 1960's (Law *et al.*, 1970), probably encouraged the earliest attempts to treat sanitary wastewater by the method. Recent symposia (Aly *et al.*, 1979; USEPA, 1980b; USEPA, 1981) and other publications (Thomas, 1979) have been essential in disseminating state-of-the-art information on the overland flow methodology.

There are many variables to be defined for a given overland flow project. These variables include:

1. pretreatment requirements
2. treatment requirements
3. storage requirements
4. soil type
5. field preparation
6. vegetation
7. wastewater application rate
8. method of application
9. on-off schedule
10. length of field
11. degree of slope
12. winter operations
13. harvesting.

Minimal pretreatment includes settling or comminution to eliminate large solids. Overland flow treatment of such minimally pretreated raw or primary wastewater can be anticipated to produce secondary level or higher quality effluent. Several authors, however, have noted the persistence of coliform bacteria in overland flow effluent of such quality, and have consequently recommended that fecal coliforms not be used as indicator organisms in the overland flow process (Peters and Lee, 1979). If phosphorus must be reduced below 1 to 2 mg/l in the final effluent, phosphorus pretreatment or post-treatment may be necessary (Peters *et al.*, 1980; Peters and Lee, 1979).

Sufficient storage volume needs to be provided to operate at reduced weekly application during the colder days of the year (Martel *et al.*, 1980; Hinrichs *et al.*, 1980). Communities presently served with stabilization ponds would normally have ample storage capacity to add overland flow treatment without the need for additional storage.

Surface soils capable of supporting a close growing cover of water resistant grasses are potentially suitable for overland flow. For effective treatment of wastewater, however, it should not penetrate the soil to more than 30 or 60 cm at most. This requires either a heavy soil or a heavy soil unit at a very shallow depth. Excessive infiltration may prevent the formation of an effective anaerobic layer at the soil surface (Myers and Butler, 1974). Also, "excessive" infiltration in overland flow actually involves relatively small volumes of soil for adsorption and exchange reactions, compared to the soil volumes normally involved in slow rate land application.

Field preparation for overland flow requires establishment of suitable cover, preceded first by coarse grading and land planing. In overland flow, it has not been worthwhile to shortcut these field preparation steps. A few experimental overland flow projects have suffered from serious erosion or treatment shortcomings because the land surface was ineffectively graded or became rutted during seeding, or because irrigation was begun before grass cover had been adequately established (Pollock, 1980; Hall *et al.*, 1979). Seeding at three to four times the rate recommended for normal agriculture has been rewarded with close, persistent cover crops which withstand the startup rigors well. Kentucky 31 tall fescue and reed canarygrass have generally done well in the cool and warm temperature United States areas where the overland flow method has been tested.

The rate of application is usually not critical for achieving adequate treatment, and up to 20 or 25 cm/wk in most settings may not be unreasonable. The rate is based upon the annual amount of wastewater to be treated, the available land area and the climate. The rate is further adjusted to allow for winter operation and time out for harvesting.

All methods of application are directed to applying wastewater uniformly across the overland flow slopes. The methods in use include application by fixed sprinklers, rotating boom sprinklers, gated pipe, perforated pipe, and spillover from split pipe and gutter. Short gravel aprons have been used to aid in equalizing the across-slope distribution. Users claim success in terms of even distribution, erosion control, and treatment effectiveness for all of these methods.

The on-off schedule has been highly variable in experimental and full scale practice. Results at this time point to 12 to 24 hr off times between 8 to 24 hr periods of continuous irrigation during the five-day week. The off-time requirement promotes a desirable balance of conditions suitable for both aerobic and anaerobic microorganisms (USEPA, 1981).

The length of the field slopes used successfully for overland flow have varied from less than 30 m to greater than 60 m. Little, if any, improvement in treatment has been demonstrated beyond the 60 m distance. Also, under equal design application rates, the greater field length offers the greater potential for erosion because greater amounts of surface water must be passed across the field surface.

The degree of slope used in successful overland flow applications has ranged from less than 2 to around 8 percent. Inadvertent pocketing with resulting pooling and establishment of nuisance vectors may occur with slopes below 2 percent. On excessively high slopes, erosion safeguards and the treatment benefits of longer water residence time on the slope are compromised.

Winter irrigation conditions have been documented by the U.S. Army CRREL studies (Martel *et al.*, 1980) and further analyses and recommendations have been made by the USEPA (Hinrichs *et al.*, 1980) for winter irrigation programming. The winter irrigation concerns are for wastewater treatment and the conditions of the vegetation cover. Soil temperatures below 4 degrees C are found to significantly inhibit the BOD removal mechanisms; also, ammonia-nitrogen treatment may fall off to nil when soil temperatures are below 14 degrees C. There is little experience to date with the behavior of vegetation under winter irrigation with pond-stabilized wastewater.

Harvesting requires an immediately prior period of about one week to allow sufficient drying time to firm the soil surface and allow the grass to dry sufficiently. Even then, low pressure soft-tired equipment is strongly recommended to avoid making ruts and grooves. The USEPA has recently recommended that thirty days total time be set aside for two harvests, to help assure adequate pre-harvest drying time (Hinrichs *et al.*, 1980).

LITERATURE CITED

Aly, O.M., D.D. Deemer, and L.C. Gilde. 1979. Overland Flow Treatment of Municipal Wastewater Effluents. Technology Transfer Seminar Series on Land Treatment of Municipal Wastewater Effluents. U.S. Environmental Protection Agency and Technological Resources, Inc., Elmer, NJ, June, unbound ms. 26 pp.

Hall, D.H., J.E. Shelton, C.H. Lawrence, E.D. King, and R.A. Mill. 1979. Municipal Wastewater Treatment by the Overland Flow Method of Land Application. U.S. Environmental Protection Agency, EPA 600/2-79-178, August, Grant No. R-803218, R.S. Kerr Environmental Research Laboratory, Ada, OK 49301.

Hinrichs, D.J., J.A. Faisst, D.A. Pivetti, and E.D. Schroeder. 1980. Assessment of Current Information on Overland Flow Treatment of Municipal Wastewater. U.S. Environmental Protection Agency, MCD-66, May, Washington, DC 20460.

Jenkins, T.F., C.J. Martel, D.A. Gaskin, D.J. Fisk, and H.L. McKim. 1978. Performance of Overland Flow Land Treatment in Cold Climates. In Proc. State of Knowledge in

Land Treatment of Wastewater, U.S. Army CRREL, Hanover, NH 03755, August 20-25, 2:61-70.

Kadlec, R.H. and Others. 1979-81. Annual Operations Reports on the Performance of the Wetland Tertiary Treatment System at Houghton Lake, Michigan. University of Michigan Wetlands Ecosystem Research Group, University of Michigan, Ann Arbor, MI 48103 (Report of 1978 Operations, 103 pp.; Report of 1979 Operations, 77 pp.; Report of 1980 Operations, 76 pp.).

Kardos, L.T., W.E. Sopper, E.A. Myers, R.R. Parizek, and J.B. Nesbitt. 1974. Renovation of Secondary Effluent for Reuse as a Water Resource. Environmental Protection Agency, Office of Research and Development, EPA-660/2-74-016, Washington, DC 20460, 495 pp.

Law, J.P., Jr., R.E. Thomas, and L.H. Myers. 1970. Cannery Wastewater Treatment by High-Rate Spray on Grassland. J. Water Pollut. Control Fed. 42:1621-1631.

Malhotra, S.K. and E.A. Myers. 1975. Design, Operation, and Monitoring of Municipal Irrigation Systems. J. Water Pollut. Control Fed. 47:2627-2639.

Martel, C.J., T.F. Jenkins, and A.J. Palazzo. 1980. Wastewater Treatment in Cold Regions by Overland Flow. U.S. Army Cold Regions Research and Engineering Laboratory Report 80-70, Hanover, NH 03755.

McCormack, D.E. 1975. Soils and Their Potential for Safe Recycling of Wastewater and Sludges. In Proc. of Second National Conference on Complete Water Reuse. American Institute of Chemical Engineers, New York, NY, pp. 629-636.

Myers, E.A. 1973. Sprinkler Irrigation Systems: Design and Operation Criteria. In W.E. Sopper and L.T. Kardos (eds.), Recycling Treated Municipal Wastewater and Sludge Through Forest and Cropland. The Pennsylvania State University Press, University Park, PA 16802, pp. 324-333.

Myers, E.A. and R.M. Butler. 1974. Effects of Hydrologic Regime of Nutrient Removal From Wastewater Using Grass Filtration for Final Treatment. Research Publication No. 88, Institute for Research on Land and Water Resources. The Pennsylvania State University, December, 65 p.

Pair, C.H. (ed.) 1969. Sprinkler Irrigation. Third Edition. The Irrigation Association, Silver Spring, MD 20907, 444 pp.

Peters, R.E. and C.R. Lee. 1979. Field Investigations of Advanced Treatment of Lagoon Effluent by Overland Flow (Abstract). In Proc. Overland Flow for Treatment of Municipal Wastewater, Greenville, SC 29602, November 27-28, pp. 27-28.

Peters, R.E., C.R. Lee, and D.J. Bates. 1980. Field Investigations of Overland Flow Treatment of Municipal Lagoon Effluent. Environmental Laboratory, U.S. Army Engineer

Waterways Experiment Station, Vicksburg, MS, USACRREL-USEPA, Working Draft of Final Report, September, 146 pp.

Pollock, T.E. 1980. The Easley (SC) Overland Flow Facility. In Proc. of a Workshop on Overland Flow for Treatment of Municipal Wastewater. U.S. Environmental Protection Agency and Clemson University, Greenville, SC 29602, pp. 77-99.

Thomas, R.E. 1980. Overland Flow: EPA Perspectives and Pre-application Treatment. In Proc. of a Workshop on Overland Flow for Treatment of Municipal Wastewater. U.S. Environmental Protection Agency and Clemson University, Greenville, SC 29602, pp. 3-7.

Trouse, A.C., Jr. 1977. Traffic Compaction and Irrigation. Annual Technical Conference Proceedings. IA. Silver Spring, MD 20907, pp. 50-53.

USEPA. 1977. Paul's Valley, Oklahoma, Overland Flow EPA Demonstration Project. Process Design Manual for Land Treatment of Municipal Wastewater, U.S. Environmental Protection Agency Tech-Trans Program Section 7-11, pp. 7-67 through 7-71.

USEPA. 1980a. The Seminar Proceedings and Engineering Assessment of Aquaculture Systems for Wastewater Treatment, September, 1979. The University of California-Davis, U.S. Environmental Protection Agency, EPA-MCD 67, U.S. Environmental Protection Agency, Washington, DC 20460, 485 pp.

USEPA. 1980b. Overland Flow for Treatment of Municipal Wastewater. In Proc. of a Workshop on Overland Flow for Treatment of Municipal Wastewater. U.S. Environmental Protection Agency and Clemson University, Greenville, SC 29602, November 27-28, 1979.

USEPA. 1981. National Seminar on Overland Flow Technology for Municipal Wastewater, Dallas, September 16-18, 1980. U.S. Environmental Protection Agency, Office of Water Program Operations, Washington, DC 20460 (in press).

Williams, T.C. and J.C. Sutherland. 1980. Engineering, Energy, and Effectiveness Features of Michigan Wetland Tertiary Wastewater Treatment Systems. In The Seminar Proc. and Engineering Assessment of Aquaculture Systems for Wastewater Treatment, University of California-Davis, U.S. Environmental Protection Agency, EPA-MCD 67, U.S. Environmental Protection Agency, Washington, DC 20460, pp. 141-173.

CHAPTER 3

WASTEWATER CROP MANAGEMENT STUDIES IN MINNESOTA*

Robert H. Dowdy, C. Edward Clapp, Gordon C. Marten, Dennis R. Linden, and William E. Larson
U.S. Department of Agriculture
Agricultural Research
University of Minnesota
St. Paul, Minnesota 55109

INTRODUCTION

The renovation of secondary-treated municipal wastewater by land application has received widespread attention in recent years. Strict discharge laws and energy considerations have intensified the interest in land treatment. In the context of this conference, it is the objective of this paper to discuss our research program in Minnesota with respect to land application of a secondary-treated wastewater that had passed through a dual-bed polishing filter. The scope of this project is so large that only a generalized summary of appropriate findings is presented. The reader is directed to the references for a current listing of applicable publications which will detail various aspects of this research. Within this framework, specific citations of this work will not necessarily be made within the text of the paper.

MATERIALS AND METHODS

The soil at the field site was a well-drained, Typic Hapludoll (Waukegan Series). It was uniform silt loam

**Contribution from the Soil and Water Management Research Unit and the Plant Science Research Unit, Science and Education Administration, Agricultural Research, U.S. Department of Agriculture, in cooperation with the Minnesota Agricultural Experiment Station, Paper No. 11,610, Scientific Journal Series.*

material overlying neutral to calcareous outwash gravel at about 60 cm with a water table at 140 to 150 cm and the topography was level.

The experimental layout consisted of 12 blocks of land, each about 0.1 ha, designed as a 2 x 3 factorial arrangement of a completely randomized block design with two replications (Figure 1). Six blocks were cropped with corn and six with perennial forages. Within each crop were three effluent treatments [control (none), low and high].

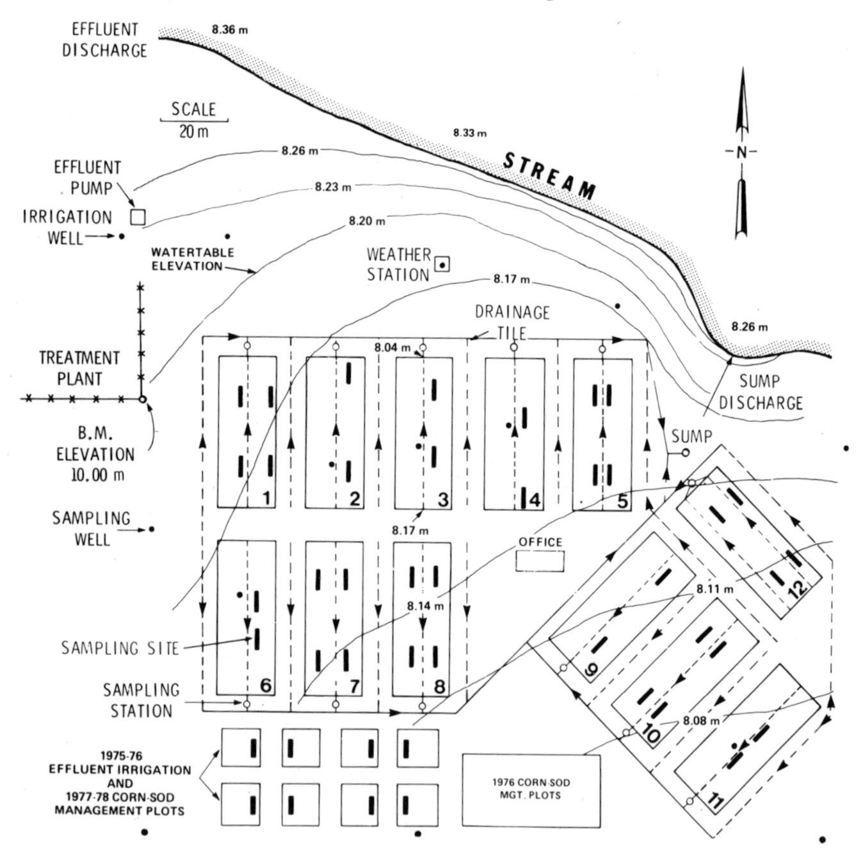

Figure 1. General plan showing tile drainage system and water sampling for Apple Valley sewage effluent project.

Effluent was sprinkler irrigated on corn and forages at rates of about 5 (low) and 10 (high) cm/wk throughout the growing season (more effluent, largely due to a longer growing season, for forage than for corn; Table I). Control treatments for both corn and forages were fertilized with mineral fertilizer and irrigated with ground water in amounts required for high dry matter production.

Table I

Annual and Total Quantities of Municipal Wastewater Effluent Applied During 4 Years of Land Application.

Crop	*Treatment*	*Year* 1	2	3	4	*Total Applied*
		cm				
Corn	Low	132	114	112	116	474
	High	214	205	219	249	887
Grass	Low	165	141	145	174	625
	High	284	246	266	318	1114

Each corn block was split in such a way that the effects of residue return or removal, tillage operations, and two different hybrids could be evaluated. Corn was planted into a strip-tilled rye cover crop to maintain adequate infiltration. Forage blocks were split to ultimately represent two-, three-, and four-time cuttings per season and further split to include eight species: alfalfa ('Agate'), reed canarygrass (Minnesota experimental), tall fescue ('Kentucky 31'), smooth bromegrass ('Fox'), timothy ('Climax'), orchardgrass ('Nordstern'), Kentucky bluegrass ('Park'), and quackgrass (Minnesota experimental). Only the mean corn and thrice-harvested reed canarygrass data will be discussed in this paper.

Dry matter production and chemical composition of various plant tissues were determined. Corn leaf samples opposite and directly below the ear were collected at silking, while other corn tissues were sampled at physiological maturity. Grass assay samples were collected from each harvest and composited on a weighted average basis.

The efficiency of the various treatments with respect to wastewater renovation was evaluated by determining the quality of soil water that percolated through the soil profile. Soil water was sampled weekly by using porous ceramic samplers installed at soil depths nominally 60 cm below the surface (just above the silt loam/gravel interface) and at 15 cm above the water table (identified nominally as the 125 cm depth) at 36 sites (Figure 1). Duplicate samplers were installed at each depth. Chemical analyses of water and effluent samples included total nitrogen and phosphorus as well as ammonium, nitrate, phosphate, chloride, and various cations.

RESULTS AND DISCUSSION

The information presented in this paper covers the first four years of the research. Data represent the duration of the perennial forage studies and much of the corn studies.

Wastewater Composition

The effluent was obtained from an activated sludge treatment facility that served a domestic suburban population. The nitrogen concentration ranged from 17 to 27 mg/l, with a mean of 20.8 mg/l (Table II). However, of greater importance is the fact that approximately 80 percent of the total nitrogen was in the ammonium form, averaging 16.5 mg/l, while the nitrate content averaged 1.5 mg/l. Total phosphorus concentrations decreased from an initial level of 11.7 to 4.1 mg/l during the fourth year, and most of the phosphorus was in the phosphate form. Effluent pH was close to 8.2 every year.

Annual applications of nitrogen and phosphorus varied from year to year as a result of different nutrient concentrations and effluent applications (Table III). Although considerable variation exists within the data, approximately equal amounts of nitrogen were applied to the control and low effluent treatments with close to twice as much nitrogen applied to the high effluent areas. Annual nitrogen applications on the high corn treatments ranged from 346 to 568 kg/ha with an accumulated nitrogen addition of 1784 kg/ha over the four year period. At least 40 percent more nitrogen was applied to the grass plots. Since effluent applications were guided by the nitrogen needs for the crop, excessive additions of phosphorus were made to both corn and grass treatments. Total accumulated phosphorus additions ranged up to a high of 915 kg/ha on the high effluent grass plots. Supplemental potassium was applied to all treatments the second and succeeding years due to a potassium deficiency in both corn and grass the first year.

Dry Matter Production

Dry matter yields of corn fodder varied from year to year on all treatments (Table IV). The control yielded an average of 16.0 tonne/ha, while the low and high effluent treatments produced an average of 12.7 and 13.4 tonne/ha of dry matter, respectively. Corn grain yields followed the same trends that were observed for fodder production and accounted for 9.0, 6.8, and 7.5 tonne/ha, of the total fodder on the control, low and high treatments, respectively. Since under this experimental design, corn variety, residue return versus removal,

Table II

Chemical Composition of Municipal Wastewater Effluent Applied During 4 Years of Land Application.

Characteristic	*Year* 1	2	3	4	*Mean*		*Pooled Standard Deviation*
	---mg/l---						
Total N	27.1	17.1	18.2	21.0	20.8	±	6.3
NH_4-N	20.8	13.8	15.1	16.4	16.5	±	5.7
NO_3-N	0.8	1.3	2.3	1.7	1.5	±	1.5
Total P	11.7	8.7	9.3	4.1	8.4	±	3.1
PO_4-P	10.6	7.9	8.3	3.8	7.6	±	2.6
Na	280	216	295	304	274	±	46
Ca	70	70	67	76	71	±	11
Mg	24	24	24	24	24	±	4
K	14	11	13	11	12	±	2
Cl	438	319	415	456	407	±	80
COD	70	35	41	50	49	±	19
Suspended Solids	54	7	8	12	20	±	13
Volatile Solids	34	6	6	6	13	±	5
EC (μmhos/cm)	2040	1700	2000	2080	1960	±	260
pH	8.05	8.20	8.25	8.10	--		--

Table III

Annual and Accumulated Nitrogen and Phosphorus Application on Corn and Reed Canarygrass.

Crop	Treatment	Element	Year 1	2	3	4	Total Applied
			----------	----------	kg/ha	----------	----------
Corn	Control	N	224	336	348	336	1244
		P	135	168	0	0	304
	Low	N	344	190	203	221	958
		P	146	99	104	40	389
	High	N	568	346	398	472	1784
		P	239	179	206	86	710
Grass	Control	N	224	336	336	336	1232
		P	135	168	0	0	304
	Low	N	475	246	267	338	1326
		P	191	125	138	66	520
	High	N	831	426	485	618	2360
		P	325	214	253	123	915

tillage versus no tillage, and cover crop clipping versus no clipping between seeded till strips did not consistently influence corn yields or quality, nor cause any interactions, only mean yield values are reported.

Forage production, as exemplified by reed canarygrass yields, was not significantly reduced by effluent applications as occurred for corn production (Table IV). In fact, average reed canarygrass yields were higher on the high effluent treatment than on the control treatment (11.1 vs. 10.2 tonne dry matter/ha). However, mean grass yields were 17 percent less than mean corn fodder yields when high effluent was applied during the four year period. The reduced reed canarygrass yields in the fourth year could be attributed to a 30 percent quackgrass invasion compared to <5 percent quackgrass in previous years.

Total nitrogen uptake by reed canarygrass was consistently higher than that by corn irrigated with effluent (Table IV). Mean nitrogen uptake by corn was only 56 percent that by grass (70, 56, and 48% in control, low and high effluent treatments, respectively). These data clearly demonstrate the superior performance of reed canarygrass over corn with respect to nitrogen renovation of wastewater. However, if factors such as the economic worth of the crop are important, satisfactory forage yields of both corn and reed canarygrass can be obtained in such systems.

The persistence and performance of other forages in our study were less impressive. Alfalfa, smooth bromegrass, and timothy did not persist beyond the second year of high effluent application. Kentucky bluegrass, tall fescue, and orchardgrass did not yield as well as reed canarygrass under the two-cutting sequence on the high effluent treatments. Quackgrass did not persist satisfactorily under the low effluent treatment. Hence, these data illustrate that reed canarygrass is best adapted to wastewater utilization systems by always accumulating more nitrogen and being either the highest yielding or among the highest yielding species in all treatments.

Water Quality

Water that percolated through the soil profile was sampled by porous ceramic cups and was the "measurement of choice" for assessing the effectiveness of our land treatment system. Data from groundwater taken from shallow wells and tile drainage systems are very difficult to interpret because of the dilution from surrounding groundwater moving through the experimental site.

The concentration of inorganic nitrogen in soil water was lower under grass cropping than under corn cropping

Table IV

*Yield of Dry Matter and Nitrogen Uptake of Corn and Weed-Free Reed Canarygrass During 4 Years of Treatment With Ammonium Nitrate (Control) Compared to Two Levels of Wastewater Effluent.**

Crop	Tissue	Treatment	Year 1	Year 2	Year 3	Year 4	Mean
			metric tons/ha				
			Dry Matter				
Corn	Fodder	Control	14.1a	19.9a	16.7a	13.3a	16.0
		Low	13.4a	13.8b	12.0b	11.7a	12.7
		High	12.4a	14.2b	15.3a	11.9a	13.4
	Grain	Control	6.4a	10.7a	10.2a	8.5a	9.0
		Low	5.8a	6.9b	7.2b	7.4a	6.8
		High	5.2a	7.5b	9.6a	7.6a	7.5
Grass	----	Control	10.1a	11.4a	11.4ab	8.1a	10.2
		Low	10.5a	9.8a	9.6b	7.7a	9.4
		High	11.2a	11.5a	13.3a	8.4	11.1

			Nitrogen Uptake				
Corn	Fodder	Control	0.18a	0.26a	0.19a	0.16a	0.19
		Low	0.18a	0.14b	0.11a	0.13a	0.14
		High	0.16a	0.16b	0.16a	0.14a	0.16
Grass	----	Control	0.32a	0.29ab	0.27ab	0.21ab	0.27
		Low	0.35a	0.22b	0.24b	0.19b	0.26
		High	0.42a	0.30a	0.37a	0.26a	0.34

*Values within columns within tissues followed by different letters are significantly different ($P \leq 0.05$); Bonferroni's multiple comparison test as modified by Dunn.

(Table V). Also, the nitrogen concentration of soil water from the control plots was higher than that from either the low or high effluent treatments for both corn and grass. The higher nitrogen concentrations in the control treatment resulted from: (1) less water percolated through the soil profile, and (2) less even distribution of nitrogen applications throughout the growing season. The nitrogen concentration of soil water was <10 mg/l under grass cropping for all 4 years and years 3 and 4 for corn regardless of effluent rate or sampling depth. The leachable nitrogen was very mobile as indicated by the small difference or lack of difference in the nitrogen concentration of soil water at the 60 and 125 cm depths. In addition, 2 to 9 times more nitrogen was in the soil water from the high effluent treatments than from the low effluent treatments. This observation was true for both corn and grass and was a direct result of the approximate doubling of the nitrogen application on the high effluent treatments versus low treatments. Additionally, the rates of nitrogen application for the low treatments were much closer to the nitrogen needs of the crops.

In a separate 2 year study, it was demonstrated that the application of 10 cm of effluent per week on reed canarygrass sod (either on 1 day, split between 2 days, or split between 5 days) did not affect dry matter production, the nitrogen uptake, or the nitrogen concentration of percolating soil water. In fact, no flush of nitrogen was observed during or immediately following irrigation even though an estimated one-half of the added water passed through the soil profile and into the groundwater. This could be due to the fact that a well-adapted grass species was used and that ammonium was the principal form of nitrogen in the wastewater.

Phosphorus removal from the applied wastewater was essentially complete. The phosphorus concentrations in soil water of the corn treatments were always <0.1 mg/l at both the 60 and 125 cm depth in spite of an accumulated addition of 710 kg/ha of phosphorus on the high effluent treatment over the 4 year period. Comparable phosphorus levels in the grass sod areas were ≤0.1 mg/l where up to 915 kg/ha of phosphorus had been applied on the high treatment. How long the system could function without a breakthrough of phosphorus into groundwater is uncertain. However, preliminary phosphorus characterization and adsorption studies suggest only minor reductions in the adsorption capacity of this soil as a result of a mean phosphorus application of approximately 250 kg/ha/yr.

RECOMMENDATIONS

We believe our system for removing nitrogen and phosphorus from municipal wastewater by land treatment with 60 cm

Table V

Inorganic Nitrogen Concentration of Soil Water During 4 Years of Wastewater Application.

Crop	*Treatment*	*Year*	*Nitrogen Concentration* in Soil Water* 60 cm	*125 cm*
			--------mg/l--------	
Corn	Control	1	83	--
		2	115	26
		3	60	47
		4	17	25
	Low	1	26	--
		2	12	14
		3	1	3
		4	2	2
	High	1	32	--
		2	10	13
		3	4	7
		4	8	8
Grass	Control	1	8.5	--
		2	7.7	5.5
		3	0.6	0.8
		4	7.0	0.5
	Low	1	7.0	--
		2	0.4	0.2
		3	0.2	0.1
		4	0.4	0.3
	High	1	12.0	--
		2	2.8	1.6
		3	1.4	2.1
		4	5.0	5.6

*Sum of ammonium- and nitrate-nitrogen as weighted mean concentration.

of silt loam soil was successful and practical. Within the boundary conditions of this study, the data summarized clearly demonstrate that reed canarygrass or corn can effectively utilize the nitrogen associated with wastewater for dry matter production. Grasses have a longer growing season for nitrogen uptake, but also yield lower economic return than does corn in areas where corn can be efficiently grown.

An extrapolation of our findings to other situations requires considerable thought, because of the many soil, climate, and wastewater characteristics that are specific for a given site. Because these factors are interdependent, drawing meaningful recommendations from this study for an undefined specific site requires special attention. Also, a high degree of management skills and expertise are required for the successful design and operation of any system for land treatment of wastewater.

ACKNOWLEDGMENT

Appreciation is expressed to the U.S. Army Cold Regions Research and Engineering Laboratory, Hanover, NH 03755, for partial support of this research.

REFERENCES

Clapp, C.E., D.R. Linden, W.E. Larson, G.C. Marten, and J.R. Nylund. 1977. Nitrogen Removal From Municipal Wastewater Effluent by a Crop Irrigation System. In R.C. Loehr (ed.), Land as a Waste Management Alternative, 8th Annual Cornell Agric. Waste Management Conf., Ithaca, NY, pp. 139-150.

Clapp, C.E., A.J. Palazzo, W.E. Larson, G.C. Marten, and D.R. Linden. 1978. Uptake of Nutrients by Plants Irrigated With Municipal Wastewater Effluent. In Proc. State of Knowledge in Land Treatment of Wastewater, Hanover, NH 03755, August 20-25, 1:395-404.

Dowdy, R.H., G.C. Marten, C.E. Clapp, and W.E. Larson. 1978. Heavy Metals Content and Mineral Nutrition of Corn and Perennial Grasses Irrigated With Municipal Wastewater. In Proc. State of Knowledge in Land Treatment of Wastewater, Hanover, NH 03755, August 20-25, 2:175-181.

Gupta, S.C., M.J. Shaffer, and W.E. Larson. 1978. Review of Physical/Chemical/Biological Models for Prediction of Percolate Water Quality. In Proc. State of Knowledge in Land Treatment of Wastewater, Hanover, NH 03755, August 20-25, 1:121-132.

Linden, D.R. 1977. Design, Installation and Use of Porous Ceramic Samplers for Monitoring Soil Water Quality. USDA Tech. Bull. 1562. 11 pp.

Linden, D.R., C.E. Clapp, and J.R. Gilley. 1981. Effects of Municipal Wastewater Effluent Irrigation Scheduling on Nitrogen Renovation, Reed Canarygrass Production and Soil Water Conditions. J. Environ. Qual. (submitted).

Linden, D.R., W.E. Larson, R.E. Larson, and C.E. Clapp. 1978. Agricultural Practices Associated With Land Treatment of Domestic Wastewater. In Proc. State of Knowledge in Land Treatment of Wastewater, Hanover, NH 03755, August 20-25, 2:183-190.

Marten, G.C., C.E. Clapp, and W.E. Larson. 1979. Effects of Municipal Wastewater Effluent and Cutting Management on Persistence and Yield of Eight Perennial Forages. Agron. J. 71:650-658.

Marten, G.C., R.H. Dowdy, W.E. Larson, and C.E. Clapp. 1978. Feed Quality of Forages Irrigated With Municipal Sewage Effluent. In Proc. State of Knowledge in Land Treatment of Wastewater, Hanover, NH 03755, August 20-25, 2:183-190.

Marten, G.C., W.E. Larson, and C.E. Clapp. 1980. Effects of Municipal Wastewater Effluent on Performance and Feed Quality of Maize vs. Reed Canarygrass. J. Environ. Qual. 9:137-141.

Marten, G.C., D.R. Linden, W.E. Larson, and C.E. Clapp. 1981. Maize Culture in Reed Canarygrass Sod to Renovate Municipal Wastewater Effluent. Agron. J. 73:(in press).

Nylund, J.R., R.E. Larson, C.E. Clapp, D.R. Linden, and W.E. Larson. 1978. Engineering Aspects of an Experimental System for Land Renovation of Secondary Effluent. Special Report 78-23. U.S. Army Cold Regions Research and Engineering Laboratory, Hanover, NH 03755. 26 pp.

USDA, AR. 1981. Development of Agricultural Management Practices for Maximum Nitrogen Removal From Municipal Wastewater Effluent. Special Report. U.S. Army Cold Regions Research and Engineering Laboratory, Hanover, NH 03755 (in press).

CHAPTER 4

CROP MANAGEMENT STUDIES AT THE MUSKEGON COUNTY MICHIGAN LAND TREATMENT SYSTEM*

Boyd G. Ellis, A. Earl Erickson,
Lee W. Jacobs, and Bernard D. Knezek
Department of Crop and Soil Sciences
Michigan State University
East Lansing, Michigan 48824

INTRODUCTION

The treatment of wastewater by passing it through agricultural crop land makes use of the "biological" filter to remove and recycle nutrients. Since the biological portion of such treatment is dependent upon crop species, soil, and the interaction between the two, living filter systems become very site specific. Certain generalizations may be made concerning different ions.

The nitrogen removal by corn and many agronomic crops is from 100 to 250 kg N/ha which is adequate to remove nitrogen from most wastewaters when applied at rates from 75 to 150 ha cm per year. But two difficulties arise. First, agronomic crops will not remove nitrogen totally from the water that is in contact with the root system. Good data are not available on this point, but the crop probably will appear nitrogen deficient if the level of nitrate in solution falls below 5 mg/l nitrogen as nitrate during the growing season. While this is higher than might be desired, it does still remove the nitrate to a level below that of the drinking water standard of 10 mg/l nitrogen. Secondly, although corn will remove considerable nitrogen, most of this is absorbed during a relatively short period of time. Aldrich (1965) states that the corn plant will accumulate more than 60 percent of its nitrogen by the eighth week after emergence. Data from Bar-Yosef and Kafkafi (1972) showed that irrigated corn accumulated more than 90 percent of its nitrogen between the fourth

**Contribution from the Crop and Soil Sciences Department, Michigan Agricultural Experiment Station Journal Article No. 9806.*

and eighth week after planting. And this was irrespective of the nitrogen fertility level.

These data strongly suggest that difficulties will arise if corn is used to remove nitrogen from wastewater. For example, at Muskegon if the nitrogen level of the water averages 10 mg/l nitrogen, 140 cm of effluent would be sufficient to produce 130 hl/ha (150 bu/acre) of corn. But that quantity of water is applied over an eight month period whereas the corn plant is accumulating 90 percent of its nitrogen during a one month period. Therefore, one can predict (and it was observed during the first year of field operation) that corn yield will be much less than 130 hl/ha due to a shortage of nitrogen during this critical stage of growth. But at the same time nitrogen will be lost to the drainage water at other periods during the year. If 8.9 cm of effluent (containing 10 mg/l nitrogen) were applied per week, one can calculate that sufficient nitrogen would be supplied for a corn yield of 33 hl/ha. Although this rationale may not be completely valid, similar corn yields were observed during the first year of operation at Muskegon when supplemental nitrogen was not added.

Although phosphorus is generally added in quantities much greater than needed by an agronomic crop, it is usually removed from the solution by the soil either through adsorption or precipitation mechanisms. Thus, the fact that application periods do not correspond to growing periods is not as critical for phosphorus as for nitrogen. The soil adsorbs phosphorus whenever it is applied, and a portion of this adsorbed phosphorus is available for plant growth when the crop is actively growing. Therefore, a wastewater renovation system must be designed so that the soil's ability to adsorb phosphorus is not exceeded by the phosphorus additions made during the anticipated life of the system.

The cations calcium, magnesium, potassium, and sodium are also supplied in wastewater. But these elements are usually present in adequate quantities and will have little effect on the crop growth. Although proportionally high sodium levels may have an adverse effect on soil structure, the soils at Muskegon are very sandy; therefore, this effect is minimal for these soils.

Heavy metals are generally low in wastewater and are primarily a concern for sludge application. No evidence has been reported which would indicate that these metals will accumulate in soils irrigated with treated sewage effluent.

While the primary Muskegon wastewater treatment system is receiving water that is low in total nitrogen, thus the problems with nitrogen management were minimal, the potential for difficulties if the nitrogen content of the water is increased made this an important question to address. The objective of this work is to evaluate cropping systems for management of nitrogen under wastewater application.

EXPERIMENTAL METHODS

The soils at the Muskegon wastewater facility are sandy and quickly react to the loading of wastewater and sludge. Leaching of nutrients is most easily studied by groundwater monitoring so a site with naturally poorly drained soils was selected. This site (circle number 26) had been artificially drained with plastic drain tile placed greater than 1.5 m deep and with a 150 m spacing between the tile lines.

The soils were a complex of AuGres sand (Entic Haplaquod, sandy mixed, mesic) and Roscommon sand (Mollic Psammaquent, mixed, mesic) with small areas of Saugatuck sand (Aeric Haplaquod, sandy, mixed, mesic, ortstein) and Croswell sand (Entic Haplorthod, sandy, mixed, mesic).

Treatments and Plot Arrangement

Treatments were designed to compare a cash crop (corn), a cash crop-forage (corn-rye or corn-oats), and forage (ryegrass) for their effectiveness in stripping nutrients, particularly nitrogen, from wastewater. Since the water table was to be sampled, the plots were 61 m square to prevent contamination from adjacent plots. Plots (in a randomized block design) were also arranged so that common boundary between rows of plots were, insofar as possible, midway between the drainage tile lines (Figure 1). This theoretically minimized the mixing of water draining from one row of plots through the other.

In 1977, a second experiment was established on the north half of the irrigation circle similar to the first but using two sludge treatments (7 and 22 tonne/ha) with corn in addition to the three primary cropping treatments. This experiment also received high nitrogen irrigation water while the initial experiment received normal wastewater. Thus, a randomized block design was used with five treatments and four replications. Only the crop management treatments will be discussed in this report.

To obtain a high nitrogen wastewater, liquid fertilizer (28% nitrogen) was injected into the wastewater whenever the irrigation rig was in the north part of the field.

Forage Crop Evaluation

An experiment (not replicated) was established in the spring of 1976 at the south end of the circle to determine

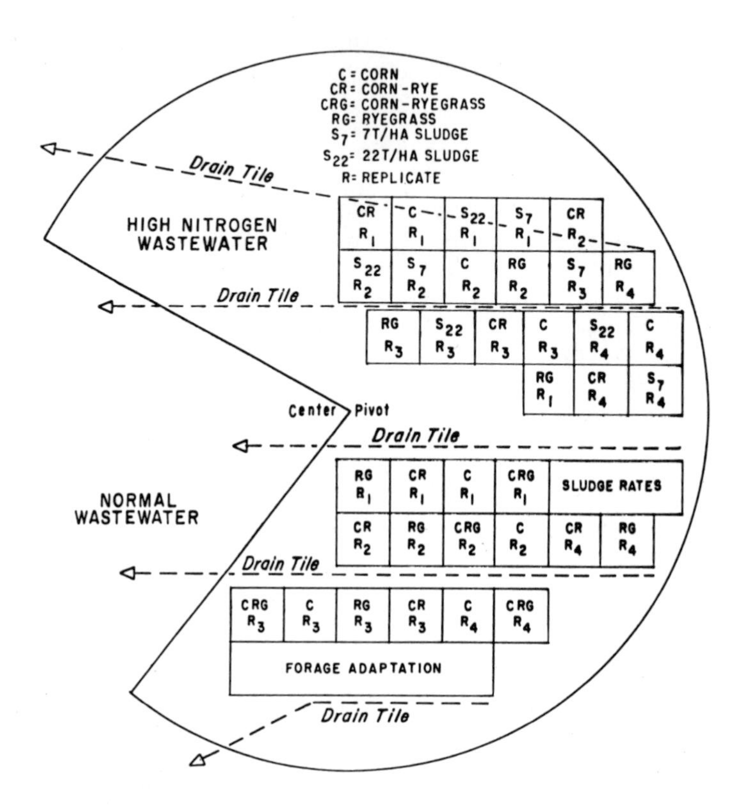

Figure 1. Plot arrangement for low and high nitrogen N-stripping experiments.

which forages were best adapted to the wastewater farm environment. They included the following crops: reed canarygrass *(Phalaris arundinacea)*, perennial ryegrass *(Lolium perenne)*, tall fescue *(Festuca arundinacea)*, orchardgrass *(Dactylis glomerata)*, alfalfa *(Medicago sativa)*, crown vetch *(Coronilla varia)*, red clover *(Trifolium pratense)*, birdsfoot trefoil *(Lotus comiculatus)*, rye *(Secale cereale)*, and quackgrass *(Agropyron repens)*.

Crop Establishment

Corn was planted in 76 cm rows with weeds controlled by atrazine for the corn treatments. Corn variety Jacques JX122A was planted in 1976; Funks SX1692 in 1977; and Pioneer 3965, Trojan, and Golden Harvest in 1978. Because the late start in the spring of 1976, substitutions were made to obtain comparative data during the first year. A mixture of Tetralite ryegrass and tall fescue was seeded in 1976 as the forage crop. This made it possible to compare a vigorously growing but late planted forage with corn in the first year. The corn-rye treatment was carried through as planned. Because there was no over-wintered rye in which to plant corn, an extra treatment, consisting of a ryegrass and oat mixture which would quickly establish itself and simulate an over-wintering forage forage-corn, was established as quickly as possible. The corn was then no-till planted in 76 cm rows in the three week old forage with a 23 cm wide paraquat treated strip over the row.

Sampling Procedures

The wastewater being applied to the fields was sampled both at the spray nozzles during the application of wastewater and by collection vessels placed within the treatment areas. Although the collection vessels gave verification of uniform application rates, the samples tended to become contaminated with dust and insects. Therefore, once comparisons indicated that the better samples were obtained from the spray nozzles, the spray nozzle data were used.

The movement of nutrients in the soil on the nitrogen stripping experiment was followed by weekly soil sampling from the 0-15, 15-30, 30-45, and 46-61 cm layers in 1976. Since the deepest samples agreed very well with the groundwater samples in 1976, only the 0-15 and 15-30 cm layers were sampled weekly in 1977 and 1978, and deep samples were taken only at the beginning and end of the season. Twenty cores were taken with a 5 cm stainless steel bucket auger and mixed thoroughly in a pail. Then a sub-sample for analysis was obtained using a stainless steel sample splitter and was placed in a plastic bag and refrigerated until analyzed.

The groundwater below each plot in the nitrogen stripping experiment was monitored for nutrient leaching. Four slotted 4 cm diameter plastic well points were placed within each plot so that the point reached into the water table. The well water sample was pumped from the upper 2.5 cm of the water table after the well had been flushed by withdrawing and discarding approximately one liter of water. Samples were stored in plastic bottles under refrigeration until analyzed.

Final corn yields were determined by hand harvesting several 15.2 m rows in each plot, drying to less than 15 percent moisture, reweighing, and calculating yields on the basis of 15 percent moisture.

Analytical Methods

Wastewater and groundwater samples were analyzed by standard methods (EPA, 1974). The analytical determinations for soils were the same as for water samples after the ions had been extracted. Detailed extraction procedures are given in Ellis *et al.* (1981).

RESULTS AND DISCUSSION

The composition and quantities of the wastewater applied are shown in Table I. The wastewater from Muskegon is lower in nutrients than most wastewaters because more than one-half of the flow comes from an industrial wastewater that is very low in nitrogen and phosphorus. Generally, total nitrogen averaged about 15 mg/l nitrogen for the low nitrogen plots and was increased to about 24 mg/l nitrogen for the high nitrogen plots.

About 83 cm of water were applied during the first year, and slightly over 100 cm were applied in each of the next two years. The total nitrogen loading for these quantities of wastewater are included in Table V and VI. Phosphorus added was the same for all treatments and was 23 to 28 kg P/ha per year.

Movement of the wastewater was verified by chloride analysis (Table II). Approximately four weeks after application of effluent, chloride reached the groundwater wells. By 1977 and 1978, the chloride content of the groundwater was approximately equal to that of the wastewater being applied showing that there was no dilution other than by rainfall. Thus, any change in the nutrient status of the wastewater could be attributed to biological removal or modification by soil chemical processes.

For simplicity, this discussion will only be about groundwater data. More complete data is given by Ellis *et al.* (1981). During 1976, the first year of the study, oats planted with ryegrass were used as the cash crop-forage (intercrop) since they could be established more quickly. The forage was planted as a mixture of Tetralite ryegrass and tall fescue but this quickly became dominated by the ryegrass. Figure 2 shows the data for the nitrate content of the groundwater in 1976. It was about four weeks after application of wastewater before nitrate began to appear in the groundwater which agrees well with the chloride data. The corn treatment and ryegrass treatment were not effective in removing nitrate during the early stages of the experiment. The intercropped treatment was almost immediately effective in reducing the level of nitrate in the water reaching the groundwater. Except for one sample (week 37) where it increased to about 10 mg/l nitrogen, this treatment effectively removed nitrogen during the growing season. The ryegrass treatment was slow to establish and did not effectively remove nitrogen until about the 31st week. But after that time it removed nitrogen

Table I

Summary of the Chemical Composition, pH, and Quantity of Applied Wastewater (Ellis et al., 1981).

Year and Nitrogen Level	*NO_3*	*NH_4*	*Total N*	*Cl*	*TP*	*K*	*Ca*	*Mg*	*pH*	*Water Applied*
	mg/l									cm
1976 Low N	5.1	3.13	14.48	240	2.30	10.9	65.5	14.4	8.11	83.5
Std. Dev.	5.4	2.04	6.05	54	0.54	1.6	10.4	1.9	0.26	
1977 Low N	4.28	6.00	16.86	280	2.31	10.9	70.2	19.5	7.73	104.1
Std. Dev.	6.11	4.09	15.20	77	0.41	2.1	8.9	7.9	0.26	
1977 High N	6.01	8.54	24.55	271	2.46	10.7	69.1	20.2	7.76	104.1
Std. Dev.	6.72	5.77	20.05	64	0.75	1.1	9.2	8.8	0.27	
1978 Low N	3.85	5.84	16.53	181	1.68	15.0	86.5	14.7	7.75	101.6
Std. Dev.	2.72	6.37	12.54	19	0.53	5.0	20.0	1.3	0.29	
1978 High N	5.80	10.25	24.25	182	1.92	16.6	86.8	14.4	7.79	101.6
Std. Dev.	2.76	7.79	17.24	19	0.36	4.5	23.4	1.5	0.19	
Low N Average	4.41	4.99	15.96	234	2.10	12.3	74.1	16.2	7.86	96.3
High N Average	5.90	9.40	24.4	226	2.19	13.6	78.0	17.3	7.78	102.8

Table II

Chloride Content of Soils and Groundwater Wells (Ellis et al., 1981).

Year	Sample		Treatment					
			Corn		Corn-Rye		Ryegrass	
			Low N	High N	Low N	High N	Low N	High N
			mg/l Cl					
1976	Week 20*	0-15 cm	4.1	---	4.0	---	3.7	---
1976	Week 20	61-91 cm	2.8	---	1.9	---	1.9	---
1976	Week 24	wells	18.6	---	38.5	---	71	---
1976	Average	0-15 cm	34.0	---	35.7	---	40.6	---
1976	Average	61-91 cm	12.9	---	14.1	---	11.6	---
1976	Average	wells	132	---	141	---	145	---
1977	Average	wells	239	199	246	210	263	206
1977	Week 46	wells	186	157	190	170	180	182
1978	Average	wells	130	135	128	143	135	133

*Denotes week of the calendar year.

effectively. Except for the period between the 34th and the 36th week, corn alone removed relatively little of the nitrogen.

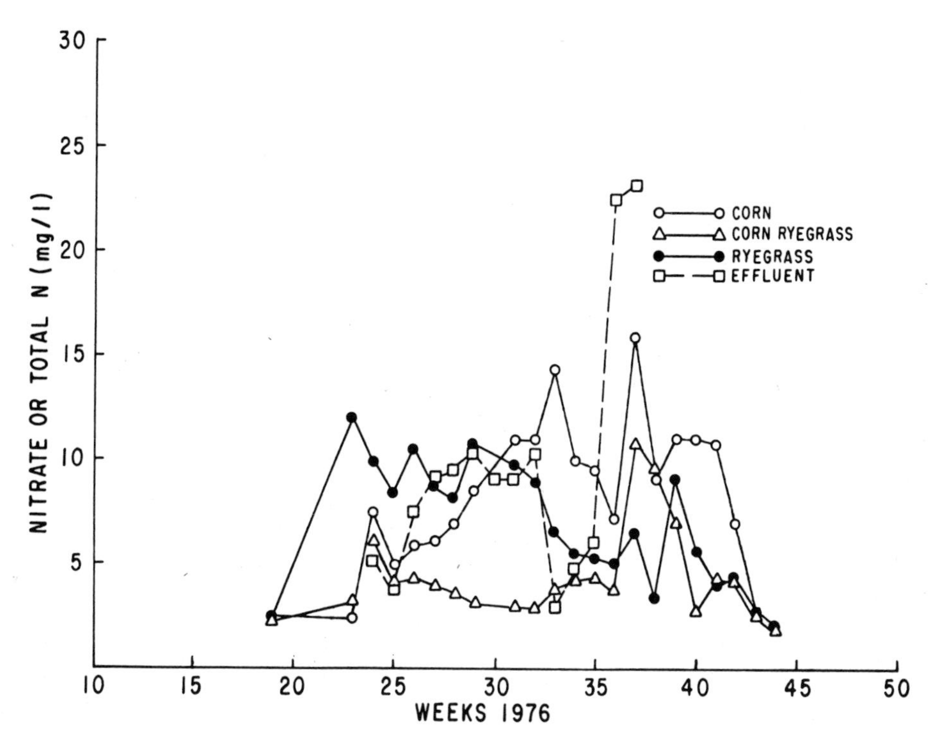

Figure 2. Mean nitrate concentration in 1976 well water samples from the surface of the water table below various crop treatments and total nitrogen concentration of the applied low nitrogen wastewater (Ellis et al., 1981).

The trends were similar in 1977 and 1978 (Figures 3, 4, and 5). But once ryegrass was well established, it was the most effective treatment in removing nitrogen from the wastewater. Corn alone was not effective until about the 34th week in all three years. Corn intercropped with rye or ryegrass was much better than corn alone in removing nitrogen, particularly in 1977. The intercrop was less effective in 1978 because the paraquat was spread on a windy day and an overkill of the rye or ryegrass occurred. The system that was used to kill a "narrow strip" of the intercrop so that corn could be established needs to be modified if used under adverse conditions. But the delicate balance between intercrop and corn that is necessary to effectively remove nitrogen from the wastewater is shown by comparing the 1977 to the 1978 data.

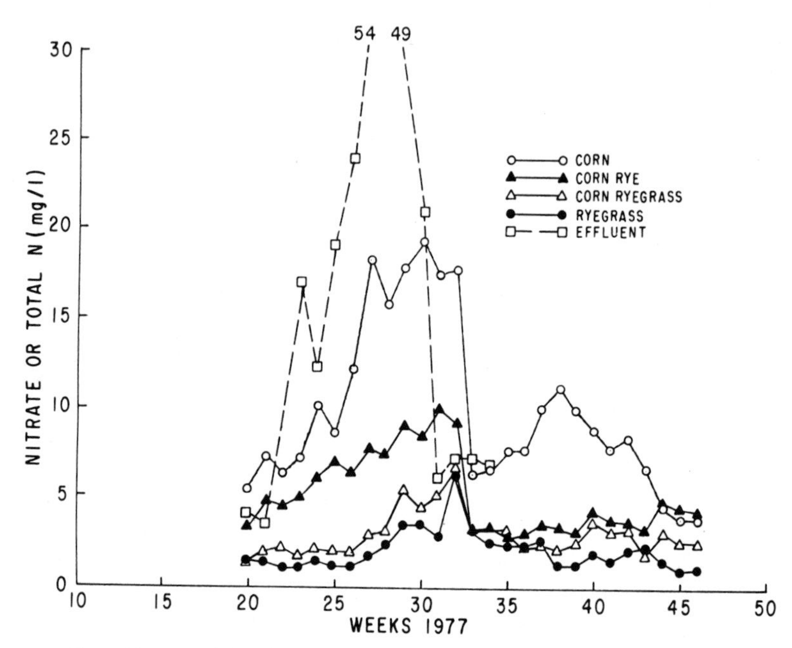

Figure 3. Mean nitrate concentration in 1977 well water samples from the surface of the water table below various crop treatments and total nitrogen concentration of the applied low nitrogen wastewater (Ellis et al., 1981).

High nitrogen plots were established in 1977 to more closely simulate wastewaters that have not been diluted with industrial wastewater. Although the data are similar to the low nitrogen plots, the concentrations of nitrate are higher. Thus, for the low nitrogen plots, even corn alone seldom left quantities of nitrate in the groundwater that exceeded the drinking water standard. But when high nitrogen water was applied (24 mg/l nitrogen), corn alone did not adequately reduce the concentration of nitrate reaching the groundwater to below the drinking water standard (Figure 3). Only during a short period of time when corn was growing rapidly (weeks 33-35) were nitrate levels effectively reduced. Ryegrass was effective in reducing the nitrate levels to an acceptable level at all times. The corn-rye intercrop was much more effective than corn alone, although for several weeks of the year the concentration in the groundwater was about 15 mg/l nitrogen as nitrate under this treatment.

The yield of corn from various treatments is given in Table III. Corn intercropped with rye generally yielded about the same as corn alone. Corn intercropped with ryegrass yielded considerably less indicating that it could not

Table III

Yields of Corn and Silage on Nitrogen Stripping Experiment (Ellis et al., 1981).

Crop	*Corn Grain*			*Corn Silage*		
		hl/ha			*Tonne/ha*	
	1976	*1977*	*1978*	*1976*	*1977*	*1978*
Low N Corn	40.6a*	68.9b	53.4a	26.0	26.3b	29.6a
Low N Corn in Rye	42.6a	71.5b	54.1a		25.7b	28.5a
Low N Corn in Ryegrass	23.5b	36.6c	49.8ab	6.4	13.7c	25.4ab
High N Corn		72.5b	35.4abc		25.3b	23.4ab
High N Corn in Rye		42.6c	21.7c		14.3c	14.5b
LSD (.05)		18.8	18.5		8.2	10.6

*Yields with the same letter are not different from others in the same column.

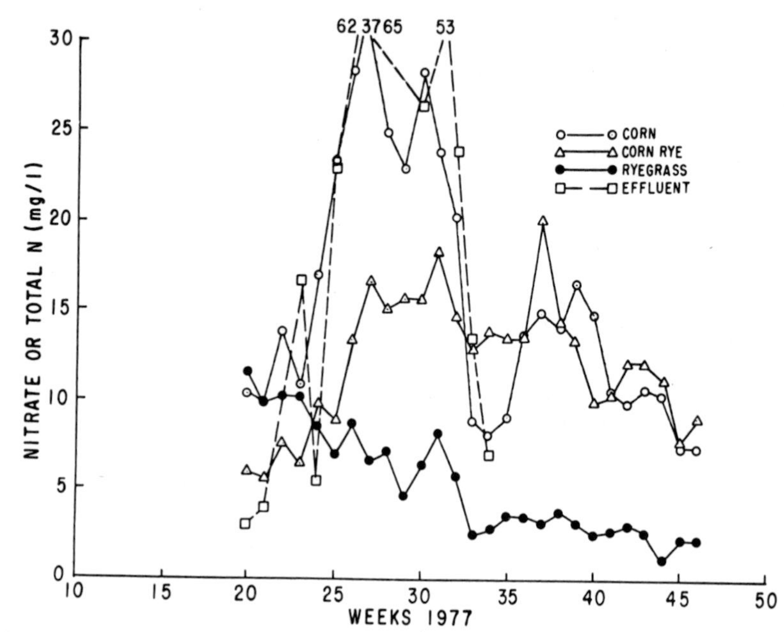

Figure 4. Mean nitrate concentration in 1977 well water samples from the surface of the water table below various crop treatments and total nitrogen concentration of the applied high nitrogen wastewater (Ellis et al., 1981).

compete effectively with ryegrass for nutrients (nitrogen probably being the most critical). The high nitrogen study produced lower yields in 1978 than the low nitrogen study because the drainage on the north portion of circle 26 was not effective. Increased rainfall during critical portions of 1978 produced flooding on this portion of the circle.

The various forages which were established in 1976 were intercropped with corn in 1978 by no-till planting corn into the various forages. The yield of the corn is given in Table IV. Tall fescue and orchardgrass competed too vigorously to be considered as an intercrop. Generally, the legume systems resulted in higher yields of corn than grass systems probably due to less competition for nitrogen.

An attempt was made to calculate a mass balance so that the nitrate leached could be compared to that added in the wastewater. The results are given in Tables V and VI for 1977 and 1978, respectively. With corn alone, about 60 percent of the added nitrogen was leached in 1977. For low nitrogen, intercropping reduced this to 32 and 17 percent for corn-rye and corn-ryegrass, respectively. For high nitrogen

Table IV

Corn Yields on the Supplementary Forage Plots in 1978 (Ellis et al., 1981).

Treatment	*Corn Yield (hl/ha)*
Red Clover	55.0a*
Alfalfa	50.1ab
Birdsfoot Trefoil	45.5ab
Ryegrass	45.1ab
Rye	37.4bc
Quackgrass	34.5bc
Reed Canarygrass	26.0cd
Tall Fescue	13.1d
Orchardgrass	13.0d
LSD (0.05) = 15.4	

*Yields with the same letter are not statistically different from each other.

wastewater, corn-oats reduced it to 44 percent lost. Ryegrass removed all but 12 percent of the added nitrogen. In 1978, the intercropped species were less effective because of the overkill of the intercrop during planting corn. Again, ryegrass hay was effective in removing nitrogen.

SUMMARY AND CONCLUSIONS

Corn alone only removes nitrogen effectively during a relatively short period of the growing season in mid-summer. Forages, such as ryegrass, are effective in removing nitrogen during a much longer period of time. However, corn can still be grown as a cash crop and the nitrogen removed by use of cover crops if an intercropping system is utilized to allow the intercrop to take up nitrogen during the period while corn is being established. Although the work reported demonstrates that this can be successfully accomplished, obtaining the correct "kill" of the intercrop is difficult but critical if one is to maintain a system that will: (1) remove nitrogen for most of the period in which wastewater is applied and (2) at the same time obtain a good yield of corn. Accomplishing this is particularly difficult if corn must be planted during adverse conditions (*i.e.*, on windy days).

Table V

Estimate of Nitrogen Loss Through Discharge to Tile During the 1977 Irrigation Season (Ellis et al., 1981).

Crop and Treatment		Total Nitrogen Added	Nitrate Nitrogen Leached*	Percent Leached
		--------kg/ha--------		
Corn	Low N	209	123	59
Corn-Rye	Low N	209	67	32
Corn-Ryegrass	Low N	209	34	17
Ryegrass Hay	Low N	209	25	12
Corn	High N	320	205	64
Corn-Oats	High N	320	142	44
Ryegrass Hay	High N	320	71	22

*Estimated loss from May 20 to November 20 based on available meterological data from Muskegon Co. Airport May 20 to July 16 and Circle 26 data (July 16 to November 20).

Assumptions:

(1) The nitrate concentration at the surface of the water table represents the concentration of nitrate leaving the root zone.

(2) There is no loss due to denitrification below the water table surface.

(3) The water moving the nitrate equals precipitation plus irrigation minus Thornthwaite evapotranspiration.

(4) There is no net change in water stored.

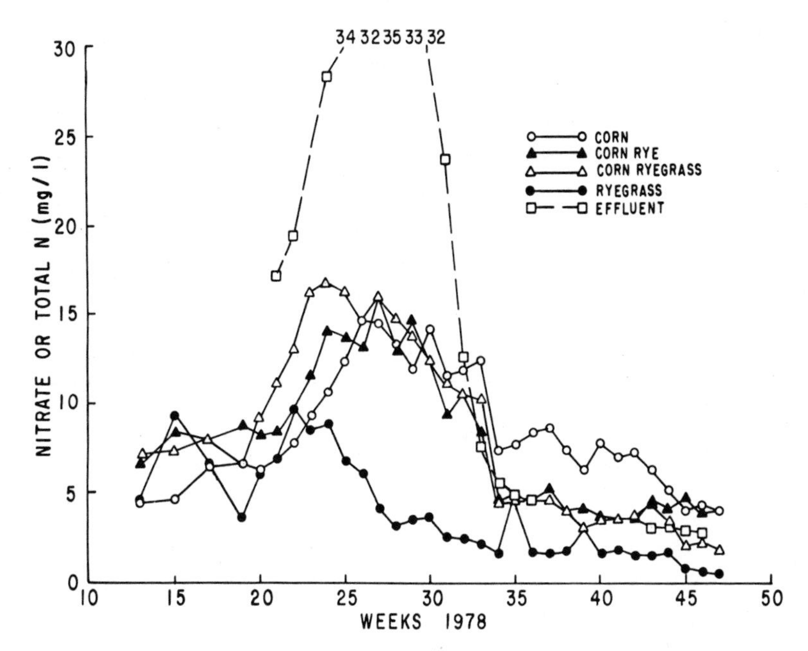

Figure 5. Mean nitrate concentration in 1978 well water samples from the surface of the water table below various crop treatments and total nitrogen concentration of the applied low nitrogen wastewater (Ellis et al., 1981).

Table VI

Estimate of Nitrogen Loss Through Discharge to Tile During the 1978 Irrigation Season (Ellis et al., 1981).

Crop and Treatment		*Total Nitrogen Added*	*Nitrate Nitrogen Leached**	*Percent Leached*
		--------*kg/ha*--------		
Corn	Low N	165	77	47
Corn-Rye	Low N	165	71	43
Corn-Ryegrass	Low N	165	70	42
Ryegrass Hay	Low N	165	30	18

*Estimated loss from May 22 to November 15 based on evaporation pan data.

ACKNOWLEDGMENT

The authors wish to acknowledge the U.S. Environmental Protection Agency for financial support of this investigation, the County of Muskegon, and Dr. Ara Demirijan for on-site cooperation and support, and the contributions of Dr. James E. Hook who coordinated the research effort for one year.

LITERATURE CITED

Aldrich, S.R. 1965. Modern Corn Production. F. & W. Publishing Corp., Cincinnati, OH, 308 pp.

Bar-Yosef, B. and U. Kafkafi. 1972. Rates of Growth and Nutrient Uptake by Irrigated Corn as Affected by N and P Fertilization. Soil Sci. Soc. Amer. Proc. 36:931-936.

Ellis, B.G., A.E. Erickson, L.W. Jacobs, J.E. Hook, and B.D. Knezek. 1981. Cropping Systems for Treatment and Utilization of Municipal Wastewater and Sludge. Completion Report for Projects G-00529201 and R-805270010. U.S. Environmental Protection Agency, Office of Research and Development, Robert S. Kerr Environemntal Research Laboratory, Ada, OK 74820, 187 pp.

EPA. 1974. Methods for Chemical Analysis of Water and Wastes. EPA-625/6-74-003. U.S. Environmental Protection Agency, Cincinnati, OH 45268, 298 pp.

CHAPTER 5

COMPARISON OF THE CROP MANAGEMENT STRATEGIES DEVELOPED FROM STUDIES AT PENNSYLVANIA STATE UNIVERSITY, UNIVERSITY OF MINNESOTA, MICHIGAN STATE UNIVERSITY, AND THE MUSKEGON COUNTY LAND TREATMENT SYSTEM

James E. Hook
Agronomy Department
Coastal Plain Experiment Station
Tifton, Georgia 31793

INTRODUCTION

In the North Central Region of the United States between 1960 and 1980, four major studies of wastewater renovation through land application to croplands were conducted. Despite differences in soils, climate, wastewater strengths, and crop studies, the responses of the soil-crop systems to wastewater irrigation were generally similar. However, the studies did build upon and support one another.

The oldest of the four was the "Living Filter" project at the Pennsylvania State University (PSU) in State College, Pennsylvania. Only the cropland irrigation systems, which began in 1963, will be considered here. Concurrent studies were begun in Minnesota and Michigan in 1974. The Minnesota studies were conducted adjacent to the Apple Valley sewage treatment plant. Again, only the effluent irrigation on crops will be considered. The Michigan study to be discussed is the "Effluent for forage crop production for livestock" study which was conducted on the site of the Michigan State University Water Quality Management Facility (WQMF) from 1974 until 1978 at East Lansing, Michigan. The last study to be considered is the "Nitrogen-Stripping" study conducted in 1976 to 1978 on circle 26 of the Muskegon County Wastewater Treatment Facility, Muskegon, Michigan. No attempt will be made to comprehensively cover all aspects of the crop production research at any of these sites.

CORN MONOCULTURE

PSU monoculture corn began in 1968 and continued through 1973. Yields in 1968 were depressed due to weed competition. From 1969 to 1971 there was a decline in silage yields from 14.6 to 7.1 tonne/ha (Hook and Kardos, 1977). There appeared to be several factors contributing to this decline. Foremost was the removal in silage harvest of more nitrogen than was applied in 1969 and 1970. Harvest removed 41 and 32 percent more nitrogen than was applied in those years with the 5 cm/wk applications (Hook and Kardos, 1977). To produce the 14.6 and 12.7 tonne/ha silage yields, nitrogen must have been mineralized from previous crop residues and soil nitrogen. After the yield had dropped to 7.1 tonne/ha in 1971, the rate of effluent application was increased 50 percent to 7.5 cm/wk for 1972 and 1973. Although this increased the total nitrogen application, yields remained low, 6.7 and 8.7 tonne/ha. Tissue test confirmed visual observations of nitrogen deficiency.

While the corn was deficient in nitrogen, soil water analyses pointed out an increase in nitrate leaching. Following winter leaching of nitrogen in the PSU corn studies, soil nitrate levels were at their lowest just before planting. Immediately after planting, wastewater irrigation was begun. Both concentration of nitrogen and volume of leachate increased. While the average nitrate in soil solution at 120 cm was 9.2 mg N/l during the winter months, it increased to 11.7 mg N/l in the months of May to September (Hook and Kardos, 1977). Peak values were 20 mg N/l. While these values were not excessive, the coincidence of nitrogen leaching while the crop grew nitrogen deficient pointed to trouble in managing monoculture corn for wastewater renovation.

Part of the reason for the inefficient early season renovation could be root distribution. With 96 cm row spacing, effective rooting would not fill the row middles early enough for efficient uptake from this dilute fertilizer source. Only a fraction of the applied nitrogen would be within reach of the corn during the lag phase of the growth curve. Deficient at that time, the plant would be in poor shape for the rapid growth phase of later weeks.

In the WQMF studies with continuous corn, Hook and Tesar (1978) examined the relation of timing of nitrogen application and leaching as compared to crop growth. First a starter fertilizer, in a band at planting, was used to alleviate the early season deficiency. This got the plant off to a good start; but, when followed by wastewater irrigation in the first few weeks, excessive nitrogen leaching occurred. On all three wastewater rate areas, a 22.4 kg N/ha starter was applied. Then 2.5, 5.0, and 7.5 cm/wk of wastewater was added. The nitrate concentrations in soil water at 1.5 cm increased sharply to levels greater than 25 mg N/l. Under leaching pressure of 5.0 to 7.5 cm/wk plus rainfall, soil water at 150 cm reflected the increase almost immediately. For the next 5 to 7 weeks the leachate contained greater than 10 mg N/l. Then with corn growth at its greatest, nitrate concentrations dropped below 10 mg N/l at 150 cm and to even lower levels at the 15 cm depth. That apparent contradiction in changes with depth was reasonable if the corn was shallow rooted. Part of the nitrogen applied in 2.5 cm of wastewater could move through a 20 cm root zone during the irrigation if antecedent moisture was high as it would be when three weekly applications were used. Subsequent sample withdrawal at the 15 cm depth occurred during 24 hr post-irrigation. During that time, nitrogen remaining in this active root zone would be removed by the plant and by denitrification. Nitrogen which had escaped the shallow roots would be subject to dilution but no further uptake. The overall course of nitrogen leaching under the management used at the WQMF was leaching of starter and wastewater nitrogen during the first 6 weeks under 5.0 and 7.5 cm/wk rates. This was followed by adequate renovation for the next 8 weeks.

The lowest wastewater rate, 2.5 cm/wk, did not cause the extreme peaks of nitrate in subsoil water. That rate was low enough to have a very low downward movement of water and, consequently, of nitrogen. However, later in the season that rate was marginal in supplying corn nitrogen needs. Nitrogen deficiency symptoms were observed, but yields were as good at that rate as with the 5.0 and 7.5 cm/wk where greater leaching occurred.

There was no clear evidence of any yield decline over the four years of effluent application even though at the lower effluent rates less nitrogen was applied than removed. In 1974, the first year, yields were 15.4, 14.5, and 12.7 tonne/ha for the 2.5, 5.0, and 7.5 cm/wk application (Tesar *et al.*, 1980). Damage from mammals ruined yield estimates in 1975 and 1976. During 1977, silage yields were 11.3, 11.4, and 13.8 tonne/ha for the three rates, respectively. Thus, there was a decline of 3 to 4 tonne/ha for the lower rates in the fourth as compared to the first year.

In the Muskegon study, corn was grown without winter cover or companion crop on two treatments. These were high and low nitrogen. Both were applied in 1.2 cm applications of wastewater by means of a center pivot oscillating back and forth in a half circle. Periods of oscillation were roughly 28 hr. Urea ammonium nitrate fertilizer was injected into both high and low nitrogen rates. The low injection occurred only during the rapid growth phase of the corn; the high nitrogen occurred through the irrigation season. Yields at both rates were similar. Although only grain was removed, the silage yields would have been 7.8, 7.9, and 8.9 tonne/ha for the low nitrogen treatments in 1976, 1977, and 1978 (Ellis *et al.*, 1981). The high nitrogen treatment did not increase yields. They were 7.6 and 7.0 tonne/ha in 1977 and 1978.

The lack of yield differences suggest that the additional nitrogen, added when the corn growth was slow, was unused. It should be subject to nitrogen leaching. Unfortunately, limits imposed by center pivot operation placed all high nitrogen rates on one side of the field, the low on the other. Differences in drainage may have prevented true treatment difference between rates. In 1977, the concentration of nitrate at the surface of the water table increased to about 20 mg N/l under the low nitrogen rate before dropping. In the high nitrogen rate, those nitrate values were greater than 25 mg N/l for several weeks but they decreased to 8 to 15 mg N/l during the end of the corn growth. In 1978, failure of tile drains on the high nitrogen locations brought the water table near the soil surface, and the nitrate concentrations below the corn remained below 15 mg N/l throughout the growing season.

In both high and low nitrogen rates, the response of nitrate in the groundwater was directly related to and almost immediately coincided with changes in nitrogen content of the effluent. When nitrogen was injected for aiding corn growth, the nitrate increased sharply in the groundwater. The nitrate concentration did not decline until after nitrogen injection stopped. Although the injection was timed to coincide with the corn uptake needs, excessive nitrate leaching occurred.

In the Minnesota studies, corn was grown at the same location for 6 years with sustained high yields. Yields of silage were 13.5, 13.8, 12.0, 11.7, 12.8, and 7.4 tonne/ha dry matter for the low wastewater rate in the years from 1974 (Larson, 1979). Under the double rate of wastewater application yields of 12.4, 14.2, 15.3, 11.9, 11.8, and 7.9 tonne/ha were obtained for those years. Increasing the application of wastewater did not affect silage (grain plus fodder) yields, and grain yields were only 6 percent higher over the years. Nitrogen removals where grain and fodder was removed were 15 percent greater under the high effluent rate indicating a higher nitrogen content of the silage at that time.

The rates of effluent and nitrogen applied in the Minnesota studies provide explanation for part of the high yields. At 5 cm/wk, the low rate applied from 184 to 346 kg N/ha/yr, an average of 265 kg N/ha/yr. At 10 cm/wk, the high rate of wastewater applied 337 to 629 kg N/ha/yr with an average annual rate of 505 kg/ha. In comparison, the 5 cm/wk applications in the PSU studies added an average of 170 kg N/ha annually and in the MSU WQMF site, an average of 120 kg N/ha. With the high rates of nitrogen added to corn in the Minnesota studies, soil water nitrogen concentrations were greater, on the average, than were observed at either PSU or the WQMF. Renovation was, however, very effective, especially in view of rates applied. As at the WQMF, the Minnesota studies observed responses to the rapid growth and uptake period of corn in a lowering of inorganic nitrogen in concentrating soil water during those uptake months. This was especially evident in the high effluent rate. Mean monthly concentrations were greater than 10 mg N/l in April and May and then again in September and October (Larson, 1979; Linden *et al.*, 1981). Those results suggest that high rates of effluent could be managed successfully during the six to eight weeks of active corn growth. At other times, rates would have to be lower to prevent nitrogen leaching.

In the Minnesota studies, there was considerable effort to optimize corn residue, winter cover, and tillage for corn growth under wastewater application. Removal of the corn fodder, after the grain harvest, had little effect on yield or nitrogen leaching. Early studies there pointed out the reduction of infiltration that occurred when field preparation removed or incorporated residues. Infiltration rates by mid-season were 0.1 cm/hr under the tilled treatments, but were 1.2 to 3.7 cm/hr when rye planted for winter cover was left (Larson, 1975, 1976). With lower infiltration, there was little reason to continue tillage practices; and in subsequent years, field preparation left some residue to maintain higher infiltration.

CORN-SOD INTERCROP

In retrospect, the PSU continuous corn suffered from lowered infiltration as a result of annual tillage and lack of winter cover. Runoff was observed, especially in winter months. Soil crusting occurred annually under the irrigations. It was in response to those conditions that perennial forages were planted for winter cover following the corn harvest in 1973. Corn was planted no-till into the grasses in 1974; the results merely pointed to problems which had to be overcome. The no-till planter failed to cover the seed in the moist soil. A poor stand resulted. Herbicide control of the grasses was not effective enough, and grasses and weeds competed with the remaining corn.

A new study was established for 1975 and 1976. Six grasses: reed canarygrass, orchardgrass, tall fescue, smooth brome, Kentucky bluegrass, and timothy; two legumes: birdsfoot trefoil and crown vetch were established. There was also a no-forage area. The intent of the planting was to have a perennial forage as a winter cover which could receive effluent prior to corn planting. This would be set back with selected herbicides to allow establishment of no-till corn. By the time the forages were recovering, shading by corn should hold back the grasses until the corn matured. The forages would be living during fall and could again receive wastewater. The benefits of such a system would be increased infiltration, elimination of runoff or at least of erosion, extension of the wastewater application season, and retention of nitrogen mineralized from corn residues.

A single herbicide treatment was used in 1975 with mixed results. The more vigorous grasses such as reed canarygrass, orchardgrass, and tall fescue survived too well and provided stiff competition for the corn. The other grasses were nearly killed by the herbicide and one year's competition by the corn. The legumes which were just planted proved too weak to provide any competition. However, by fall, after corn silage was removed, the effective cover that was desired was there. The poorer grasses had given way to volunteer grasses and weeds. The legumes were well established. The vigorous grasses and the legumes were ready for the next year's management. Unfortunately, the yields and nitrogen renovation of this phase of the PSU studies have not been published.

A winter cover of rye was used in the WQMF. Besides providing erosion control, it was used to extend the irrigation season prior to corn planting. There was no effort made to retain a rye cover or rye intercrop in those studies. The rye was killed with paraquat at planting.

In the Muskegon study, use of an intercrop was included from the beginning in 1976. Both the perennial ryegrass and the annual rye were included in the comparison with corn planted without cover or intercrop. The rye was seeded aerially in August in the maturing corn. The following year corn was planted no-till into the rye. In 1978, paraquat was sprayed in a band approximately 12 inches wide over the row. The intent was to leave sufficient living rye in the interrow to allow nitrogen immobilization while the corn was becoming established. The herbicide was used to prevent rye competition with the emerging corn. Then the rye would mature as the corn was growing most rapidly. All of the nitrogen applied at that time should be available to the corn. The results of this treatment in 1977 and 1978 proved the rye could be established in the maturing corn, and the corn could be established in the maturing rye. Yields of corn were as high in this corn-rye system as in the corn alone (Ellis *et al.*, 1981).

Removal of nitrogen by the living rye in 1977 lowered nitrate concentrations in the groundwater to 10 mg N/l during the peak for the low nitrogen treatment. It lowered nitrate to 17 mg N/l during the peak for the high nitrogen treatment. This compares to peak values of 20 and 30 mg N/l for the corn alone at the low and high nitrogen treatments. In 1978, the herbicide treatment killed most of the rye. As a result, nitrogen leaching under the corn-rye was very similar to nitrogen leaching under the corn alone.

This same system was examined more closely in Minnesota. In split plots, the effect of living versus killed cover was compared. Rye was established in the corn each fall. At planting, a strip of the cover was rototilled and corn was planted into it. On one half the area, rye was cut over the whole plot. On the other, rye grew in the interrow. The effect of this management on nitrogen leaching and on yield was very small and nonsignificant (Larson, 1976).

While use of annual rye as a cover and an intercrop proved successful in both Muskegon and Minnesota studies, corn growth suffered when perennial forages were used. In the case of ryegrass at Muskegon, competition reduced corn yields 28 percent on the average to 5.9 tonne/ha (Ellis *et al.*, 1981). The reduction was due in part to failure to control the aggressive ryegrass with herbicides. It was also related to failure to control the broad spectrum of weeds in the interrow. By the third year of this no-till system, even trees were becoming a problem.

As might be expected, the ryegrass cover proved to be effective in controlling nitrogen leaching when it competed most strongly. Peak nitrate concentrations in groundwater were 6 mg N/l compared to 20 mg N/l for corn alone in 1977 (Ellis *et al.*, 1981). In 1978, when more of the ryegrass was killed by the herbicide, the corn-ryegrass was no more effective than the corn alone in controlling leaching.

In 1978, a special study was done at Muskegon to evaluate the suitability of other perennials as an intercrop in a system such as the ryegrass-corn. Corn was planted no-till into nine forages using five herbicide treatments. Those forages included reed canarygrass, alfalfa, tall fescue, orchardgrass, ryegrass, quackgrass, rye, red clover, and birdsfoot trefoil. All except the rye had been established in 1976. The purpose was to find herbicide treatments which could slow down forage growth or control it into interrow strips which could survive and regrow. Red clover, alfalfa, birdsfoot trefoil, and ryegrass were most suitable and allowed the highest corn yields. Rye and quackgrass were more competitive over all. Reed canarygrass, tall fescue, and orchardgrass had such dense sods the no-till planter could not effectively plant in the sod. Yields of corn were severely limited on those latter forages (Ellis *et al.*, 1981).

At Minnesota during the period between 1976 and 1978, reed canarygrass sods were planted with corn to develop methods of controlling competition of the grass while leaving enough grass to post-corn recovery. In one test, corn was planted in rototilled strips and the grass was controlled with various herbicide treatments. With 70 percent or more of the grass suppressed, yields of silage were high, 16.6 to 17.2 tonne/ha (Marten *et al.*, 1981). However, where the grass was killed, 95 to 100 percent, there was no recovery. Where grass control was poorer, corn yields were lower, but grass recovery was greater. Nitrogen in leachate below the root zone was lowest where grass growth was greatest.

In a second test, corn was established as before in a reed canarygrass which had received effluent at high rates for two years. Again, suppression of 60 percent or more gave the highest yields of corn (Marten *et al.*, 1981). Again, where reed canarygrass competed enough to lower corn yields, the nitrogen in soil water was lower. Selected herbicide treatments could indeed control the perennial grass enough to get good corn growth, and at the same time could allow for recovery of the grass. However, where the grass was controlled enough to prevent competition, nitrogen leaching occurred as though no grass was present.

The wide interest in corn as a marketable by-product of wastewater renovation led all four research efforts to extensive study of corn systems. At lower rates, renovation was good but yields were not. At higher rates, nitrogen leaching became a problem in spring and fall. Intercropping helped, but simple solutions to obtaining sustained high yields and preventing nitrogen leaching have not been found. What has been found is that several forage grasses are well adapted and competitive, and they provide excellent safeguards for minimizing nitrogen leaching.

PERENNIAL FORAGES - STAND SURVIVAL

Forages have been a part of each renovation study from their beginnings. In the PSU studies, red clover and alfalfa were used in a strip crop rotation involving corn (Kardos *et al.*, 1974). The rotation of corn, small grains, one or more years of hay, and the use of contour strips were recommended best management practices for farms in Pennsylvania. The practices were incorporated into the wastewater renovation study at its inception. The small grains proved ineffective and unsuitable for wastewater renovation. They were dropped in 1965. Alfalfa prevented nitrogen leaching effectively and concentrations of mineral nitrogen in soil water remained below 10 mg/l. However, because alfalfa is capable of producing its own nitrogen, and because it was giving way to the more vigorous grasses and weeds, it too was eliminated after 1967.

Reed canarygrass was established in 1963 and irrigated with wastewater year round for the next 14 years. Its selection for the wastewater system was based on its ability to survive in poorly drained soils. The grass quickly proved its suitability. It has existed as a monoculture with no management except wastewater additions and hay removal. For the years prior to sludge, additional dry matter yields had ranged from

9.7 to 15.7 tonne/ha/yr in a total of three cuttings (Hook and Kardos, 1977). The grass withstood the ice cover from winter spray applications without problems.

Removal of nitrogen from the wastewater was excellent. Since the crop itself could take up 300 to 400 kg N/ha/yr, effluent application rates could be quite high (Hook and Kardos, 1977). Through the years prior to sludge application, a single 5 cm application per week was used. Mineral nitrogen at 120 cm in soil water remained below 6 mg N/l continuously. With total annual effluent applications of 550 to 700 kg N/ha, the grass was not limited in growth by nitrogen. At the same time, total annual leaching losses were below 23 percent of nitrogen added. The results suggested that the system could have been operated continuously.

Reed canarygrass does have one drawback, however. It is considered a low quality feed due to its alkaloid content and to lower digestibility. In the Minnesota and Michigan WQMF studies, several grasses and legumes were studied in comparisons with each other and with corn. At Muskegon, a single perennial ryegrass was studied in comparisons with corn.

In the Minnesota studies, eight perennial forages were evaluated for adaptability in wastewater renovation systems. Stand survival, quality and quantity of forage, and cutting management were evaluated on alfalfa, smooth bromegrass, orchardgrass, Kentucky bluegrass, tall fescue, timothy, reed canarygrass, and quackgrass. Alfalfa, timothy, and smooth bromegrass failed to persist well when irrigated with effluent at rates of 5 to 10 cm/wk (Marten *et al.*, 1979). In the case of alfalfa, root-rot development was primarily responsible for the decline even though the cultivar 'Agate' had root-rot resistance. After two years of the effluent irrigation, the alfalfa was essentially destroyed.

Survival of the grasses was related to cutting management. At the 5 cm/wk effluent rate, Kentucky bluegrass, timothy, and tall fescue persisted better when cutting frequency increased from two times per year to four times (Marten *et al.*, 1979). Orchardgrass, reed canarygrass, and quackgrass were affected adversely by increasing the cutting frequency. Those differences were related to the morphology of the grasses. At 10 cm/wk, effluent rate relative survival of the species changed. Quackgrass, reed canarygrass, tall fescue, Kentucky bluegrass, and orchardgrass all survived better when cut four times annually rather than twice. After five years, only those latter species were considered to be suitable in persistence and yield to warrant inclusion in a wastewater forage system.

At the WQMF, lower rates of effluent were used in a comparison of eight grasses and eight legumes. Six alfalfa cultivars were tested including 'Agate'. All survived with better than 90 percent stands in the fifth year when irrigated at the 2.5 cm/wk rate (Tesar *et al.*, 1980). As the effluent rate was increased, the survival was related to susceptibility to root disease. 'Agate' had an 87 percent legume stand at the 7.5 cm/wk rate. The other five cultivars had stands of 38 to 67 percent.

All of the grasses were cut three times annually. Tall fescue, reed canarygrass, orchardgrass, and Kentucky bluegrass persisted well with 95 percent stands in the fifth year. As in the Minnesota studies, smooth bromegrass and timothy, as well as reed foxtail, performed poorly at all rates.

Although stand survival was not evaluated in the Muskegon study, the ryegrass survived well under the two cutting management (Ellis *et al.*, 1981).

Reasons for the difference in performance of alfalfa at Minnesota and Michigan could not be determined. It is certain that under large production systems, control of perennial grasses will be necessary to maintain good stands of the alfalfa.

PERENNIAL FORAGES - YIELDS

In Michigan, where it persisted, alfalfa produced higher yields than the perennial grasses. Yields over the five years at the WQMF were 12.9, 11.7, and 10.3 tonne/ha/yr for the 2.5, 5.0, and 7.5 cm/wk rates, respectively (Tesar *et al.*, 1980). The lower values for the higher rate were due to the yield decline in the last three years as nonadapted cultivars weakened. For example, the yields averaged over cultivars in the fifth year were 12.5, 9.5, and 5.8 tonne/ha for the 2.5, 5.0, and 7.5 cm/wk rates, respectively.

'Agate' alfalfa could be compared directly between study sites. In the WQMF, first year yields were 10.2, 10.9, and 10.9 tonne/ha for the 2.5, 5.0, and 7.5 rates and were 14.4, 13.6, and 11.9 tonne/ha for the second year (Tesar *et al.*, 1980). In Minnesota, first year yields were 7.54 and 7.28 tonne/ha for 5.0 and 7.5 cm/wk rates and were 5.4 and 3.5 tonne/ha for the second year (Marten *et al.*, 1979).

The persistent grasses had more consistent yields over years and between study sites. At the WQMF, yields of tall fescue, reed canarygrass, orchardgrass, and Kentucky bluegrass irrigated at 5 cm/wk were 9.9, 8.7, 8.7, and 8.0 tonne/ha, respectively, when averaged over the first three years (Tesar *et al.*, 1980). At Minnesota, yields of those species irrigated at 5 cm/wk were 10.0, 10.0, 7.1, and 7.4 tonne/ha, respectively, for those same three years (Marten *et al.*, 1979). In comparison, yields of reed canarygrass in the first three years of the PSU study were 11.0 tonne/ha.

Response of those species to increasing effluent rates differed from the response of alfalfa. In the WQMF, grass growth was limited by nitrogen availability at the 2.5 cm/wk rate. For the highest yielding grass, tall fescue, the 5 year average yield increased from 7.3 to 9.9 to 11.0 tonne/ha as the effluent rate increased from 2.5 to 5.0 to 7.5 cm/wk. Other grasses responded similarly. In Minnesota, yield increases from increasing the effluent rate from 5 to 10 cm/wk were approximately 20 percent for the four persistent grasses (Marten *et al.*, 1979).

With five adapted grasses with high forage yields for wastewater renovation systems, selection can be based on need for the crop. Orchardgrass has better animal acceptance than reed canarygrass or tall fescue. Tall fescue and Kentucky bluegrass survives well under more frequent cutting. Reed canarygrass responds more to luxury nitrogen consumption and would provide more effective renovation of high nitrogen wastewater (Marten *et al.*, 1980).

PERENNIAL FORAGES - NITROGEN LEACHING

Nitrogen renovation of all forages in the WQMF, Muskegon, and Minnesota was as good as in the PSU studies with reed canarygrass. Concentration of nitrate in soil water under grasses rarely exceeds 10 mg N/l even under the nitrogen loading rates of more than 800 kg N/ha applied in Minnesota. Even the alfalfa proved effective in controlling nitrogen leaching where it persisted. Apparently, as was suggested in the PSU studies, nitrogen was taken from soil solution by the legume rather than from the atmosphere.

Two studies probed the renovation by reed canarygrass under various irrigation schedules. At PSU, in 1973 and 1974, effluent containing municipal sludge was applied in twice weekly applications of 2.5 cm, in twice weekly applications of 5.0 cm/wk, and in a single weekly application of 5.0 cm/wk. The concentration of nitrate in the leachate and the amount of nitrogen leaching did not differ significantly between the single versus split application of 5.0 cm/wk (Hook, 1975). The double rate increased nitrogen leaching and soil nitrate concentrations exceeded 10 mg N/l continuously. Yields were not different between single and split applications, but the yield of the grass was increased with the double rate.

In Minnesota, in 1975 and 1976, effluent was applied in a single weekly application of 10 cm, in twice weekly applications of 5.0 cm, and in five weekly applications of 2.0 cm (Linden *et al.*, 1981). The concentration of nitrogen in soil water at the 125 cm depth was slightly lower both years where the weekly application was split into 5 days. However, the differences were nonsignificant, and the mean annual concentration with the single application was only 3.4 mg N/l in the second year. Since total applications were the same, the amounts of nitrogen lost from the soil would change as the concentrations. Frequency of water application had no effect on yield of the grass.

While most wastewater renovation systems will need to remove the forage to maintain production and to assure nitrogen removal, another use for the wastewater could be on turf. Kentucky bluegrass was managed under three clipping regimes at the WQMF in 1976 and 1977 (Hook and Burton, 1979). The grass was clipped every other week and clippings left on the site. The grass was harvested three times annually with the hay removed, and the grass was left uncut. Mineral nitrogen concentrations in soil water did not differ significantly among cutting managements. They remained below 10 mg N/l throughout the study period.

SUMMARY

Although there were significant differences in soil and within limits to climate, there were strong similarities in crop responses to wastewater renovation among the four studies examined. Corn was found capable of producing reasonable grain yields when nitrogen applications in the wastewater or in supplemental fertilization was high enough. Corn, however, was effective in preventing nitrogen leaching for only a short period each year. Intercropping with rye and forages could improve the renovation, but a difficult balance between corn and forage must be managed to obtain both goals. Perennial forages were very effective in preventing nitrogen leaching. Several adapted species are available to select. Cutting management of the forage affects stand survival and production of the forage.

LITERATURE CITED

Ellis, B.B., A.E. Erickson, L.W. Jacobs, J.E. Hook, and B.D. Knezek. 1981. Cropping Systems for Treatment and Utilization of Municipal Wastewater and Sludge. Completion Report of Projects G-005292 01 and R805270010. Robert S. Kerr Environmental Research Laboratory, Office of Research and Development, U.S. Environmental Protection Agency, Ada, OK 74820, 187 pp.

Hook, J.E. 1975. Distribution and Movement of Nitrogen in Sites Used for Application of Municipal Sewage Effluent and Sludges. Ph.D. Thesis, The Pennsylvania State University, College Park, PA 16802.

Hook, J.E. and T.M. Burton. 1979. Nitrate Leaching From Sewage-Irrigated Perennials as Affected by Cutting Management. J. Environ. Qual. 8:496-502.

Hook, J.E. and L.T. Kardos. 1977. Nitrate Relationships in the Penn State "Living Filter" System. In R.C. Loehr (ed.), Land as a Waste Management Alternative. Ann Arbor Science Publishers, Inc., Ann Arbor, MI 48106, pp. 181-198.

Hook, J.E. and M.B. Tesar. 1978. Land Application to Croplands. In T.M. Burton, The Felton-Herron Creek, Mill Creek Pilot Watershed Studies. EPA-905/9-78-002, U.S. Environmental Protection Agency, Region V, Chicago, IL 60604, pp. 66-85.

Kardos, L.T., W.E. Sopper, E.A. Myers, R.R. Parizek, and J.B. Nesbitt. 1974. Renovation of Secondary Effluent for Reuse as a Water Resource. Environmental Protection Technology Series EPA-660/2-74-016, U.S. Environmental Protection Agency, Washington, DC 20460, 495 pp.

Larson, W.E. (Coordinator) 1975. Utilization of Sewage Wastes on Land. Research Progress Report. Soil and Water Management Research Unit and Plant Science Research Unit, Agricultural Research, U.S. Department of Agriculture, University of Minnesota, St. Paul, MN 55108.

Larson, W.E. (Coordinator) 1976. Utilization of Sewage Wastes on Land. Research Progress Report. Soil and Water Management Research Unit and Plant Science Research Unit, Agricultural Research, U.S. Department of Agriculture, University of Minnesota, St. Paul, MN 55108.

Larson, W.E. (Coordinator) 1979. Utilization of Sewage Wastes on Land. Research Progress Report. Soil and Water Management Research Unit and Plant Science Research Unit, Agricultural Research, U.S. Department of Agriculture, University of Minnesota, St. Paul, MN 55108.

Linden, D.R., C.E. Clapp, and J.R. Gilley. 1981. Effects of Scheduling Municipal Wastewater Effluent Irrigation of Reed Canarygrass on Nitrogen Renovation and Grass Production. J. Environ. Qual. (submitted).

Linden, D.R., C.E. Clapp, and W.E. Larson. 1981. Quality of Percolate Water After Treatment of a Municipal Wastewater Effluent by a Crop Irrigation System (in preparation).

Marten, G.C., C.E. Clapp, and W.E. Larson. 1979. Effects of Municipal Wastewater Effluent and Cutting Management on Persistence and Yield of Eight Perennial Forages. Agron. J. 71:650-658.

Marten, G.C., W.E. Larson, and C.E. Clapp. 1980. Effects of Municipal Wastewater Effluent on Performance and Feed Quality of Maize vs. Reed Canarygrass. J. Environ. Qual. 9:137-141.

Marten, G.C., D.R. Linden, W.E. Larson, and C.E. Clapp. 1981. Corn Culture in Reed Canarygrass Sod to Renovate Municipal Wastewater Effluent. Agron. J. (in press).

Tesar, M.B., J.E. Hook, and B.D. Knezek. 1980. Municipal Sewage Effluent for Forage Crop Production for Livestock. Technical Completion Report. Project A-078-MICH, Office of Water Research and Technology, U.S. Department of the Interior, Washington, DC 20240.

CHAPTER 6

MANAGEMENT STUDIES OF ANNUAL GRASSES AND PERENNIAL LEGUMES AND GRASSES AT THE MICHIGAN STATE UNIVERSITY WATER QUALITY MANAGEMENT FACILITY*

M.B. Tesar and B.D. Knezek, Department of Crop and Soil Science, Michigan State University, East Lansing, Michigan 48824; and J.E. Hook, Agronomy Department, Coastal Plain Experiment Station, Tifton, Georgia 31793

INTRODUCTION

Effluent from municipal sewage systems generally causes unnecessary and undesirable contamination of lakes and streams and eutrophication of lakes, frequently with irreversible ecological changes.

Effluent, however, has been sprayed satisfactorily on cropland and forests after being treated in lagoons or in conventional sewage treatment plants in Europe and recently in Pennsylvania and other states. Raw sewage has been used by flood irrigation on clover-grass pastures in Australia for 80 years but little information on harmful accumulations of minerals in the soil was obtained. The effluent with its nitrogen, phoshorus, and potassium provides needed nutrients for good growth of forage crops. The soil acts as an absorbent or filter for phosphorus, and to a lesser extent, nitrogen. However, little is known about the maximum capacity of soils to retain and recycle these essential nutrients under high levels of water application and still maintain satisfactory crop productivity.

Administrators in Michigan cities, probably typical of many in the United States, constantly barrage the university with unanswered questions regarding which crops to grow, the amount of water to apply, the method of harvest, the utilization of crops, and the extent of soil and water contamination when treated effluent is used for irrigation in this

*Contribution from the Crop and Soil Sciences Department. This paper will be published in the Agronomy Journal.

relatively new system of sewage disposal. Research should provide such information to municipalities rather than force them to determine, generally by non-scientific and frequently inaccurate observation, what crops should be grown, irrigation levels to follow, and the effect of nitrogen and minerals on soil water.

Utilization of forage or feed production by municipalities should have a sale value to stables with riding horses, to private owners of pleasure horses, or to livestock farms for the production of proteins such as milk or meat if satisfactory crops of high digestibility are produced. If forages of high quality and yield can be produced with high loading rates of sewage effluent, their sale value can partially defray the cost of disposal of the sewage effluent.

The objectives of this experiment were as follows:

1. to determine which perennial legume and grass forage crops, in comparison to annual grass crops, will produce high yields suitable for utilization by livestock for a period of several years without re-establishment when irrigated with high levels of sewage effluent and harvested under varying frequencies imposed to obtain maximum biomass per hectare;
2. to determine the ability of the soil and plants to remove minerals from sewage effluent applied in perennial forage crops and annual crops; and
3. to determine the effectiveness of perennial grasses, perennial legumes, and annual grasses in removing nitrogen from effluent and how the removal affects nitrates in the soil water at varying depths.

MATERIALS AND METHODS

Objective 1.

Since the project involved perennial plants, stands of perennials were established in August, 1973, a year prior to collection of yield data following irrigation with effluent. Annuals were established in May, 1974, and were re-established annually in May by planting in a fall-sown crop of rye suppressed by the herbicide paraquat in mid-May. The field plots were established on a one-hectare field at the Michigan State University Water Quality Management Facility and were irrigated with varying levels of sewage effluent from the city of East Lansing, Michigan. The soil was uniform, a member of the fine-loamy, mixed mesic family of Typic Hapludalfs developed on a silt to loam till. The soil had an initial average soil test of 2400 kg Ca, 100 kg Mg, 8 kg P, 125 kg K/ha, and a pH of 5.4 in the upper 20 cm in 1973. This was adjusted to an average pH of 7.1 by incorporation of 15 tonne/ha of lime in a split application before and after plowing to a depth of 20 cm. The soil pH increased from 7.1 to 8.5 as the depth increased from 20 to 310 cm. It was not tile drained but drained well enough to grow high yields of corn under normal Michigan conditions. The water level was 3 or more meters below the soil surface.

The following were established in three replications: (1) Eight perennial grasses established in August, 1973: smooth bromegrass *(Bromus inermis* Leyss*)*, cultivar Sac (southern); smooth bromegrass *(Bromus inermis* Lyess*)*, Canadian, (northern); orchardgrass *(Dactylis glomerata* L.*)*, cultivar Nordstern; tall fescue *(Festuca arundinacea* Schreb.*)*, cultivar Kentucky 31; timothy *(Phleum pratense* Leyss*)*, cultivar Verdant; Kentucky bluegrass *(Poa pratensis* Leyss*)*, cultivar Park; creeping foxtail *(Alopecurus arundinaceus* Poir*)*, cultivar Garrison; and reed canarygrass *(Phalaris arundinacea* L.*)*, commercial. (2) Eight perennial legumes established in August, 1973: alfalfa *(Medicago sativa* L.*)*, cultivar Saranac; alfalfa *(Medicago sativa* L.*)*, cultivar Agate, resistant to *Phytophthora megasperma*; alfalfa *(Medicago sativa* L.*)*, cultivar Vernal; alfalfa *(Medicago sativa* L.*)*, cultivar 520; alfalfa *(Medicago sativa* L.*)*, cultivar Iroquois; alfalfa *(Medicago sativa* L.*)*, cultivar Ramsey; birdsfoot trefoil *(Lotus corniculatus* L.*)*, cultivar Viking; and birdsfoot trefoil *(Lotus corniculatus* L.*)*, cultivar Carroll. (3) Three annuals planted each spring starting in 1974 and harvested the same year: corn *(Zea mays* L.*)*, cultivar Mich 575-2X (single cross hybrid); sorghum-sudangrass hybrid *(Sorghum bicolor* L. Moench x *S. sudanense* P. Stapf*)*, cultivar Pioneer 988; and forage sorghum *(Sorghum bicolor* L. Moench*)*, cultivar Pioneer 931.

The experimental design was a split-plot with three replications. Annuals and perennials were whole plots and species were sub-plots within effluent levels which were not replicated because of physical limitations imposed by the minimum of a uniform soil type available. There was a total of 171 plots, 1.8 x 9.1 m in size (3.1 x 9.1 m in annual crops planted in four rows 7.6 cm apart). Yields were determined annually during a 5 year period starting in 1974 by obtaining green field weights in a 0.91 x 8.23 m harvest area of the plot in perennial forages and annuals sown in solid stand in order to eliminate species border effect. Three harvests (early June, mid to late July, and mid to late September) were made on perennial grasses and legumes. The center two rows of forage sorghum and corn were harvested in late September to early October. Dry matter samples of 1000 g were obtained for calculation of total weed free dry matter per hectare and for chemical analyses. Stand estimations on perennial species were made during the growing season and at the termination of the experiment in 1978 to monitor changes in resistance of the species to high levels of effluent, and to determine which perennial species will persist for a period of several years under various effluent levels.

Fertilizer (nitrogen and phosphorus) was added each year to help establish annuals so was added to all treatments for uniformity. Amounts in kg/ha added as nitrogen and phosphorus to all treatments were: 1974-30N and 39P; 1975-45N and 58P; 1976, 1977, 1978-20N and 15P. Potassium fertilizer was added to all treatments at an annual rate of 300 kg/ha starting in 1975 to insure that legumes had an adequate supply for maximum growth.

Three effluent loading rates of 2.5, 5.0, and 7.5 cm/wk were applied to the plots from early May through October in increments of 2.5 cm over a 3 hr period. If 2.5 or more cm of rain were received the previous day, irrigation at the 7.5 cm/wk level was deferred until the next day to reduce runoff.

The average composition of the effluent (mg/1) for the 5 year period was as follows: organic nitrogen, 2.6; nitrate-nitrogen, 1.5; nitrite-nitrogen, 0.005; ammonia-nitrogen, 9.3; soluble phosphorus, 3.0; total phosphorus, 7.0; BOD, 135; suspended solids, 155; potassium, 8.0; chloride, 272, iron, 3.0; manganese, 0.16; zinc, 0.49; nickel, 0.23; copper, 0.13; cadmium, <0.03; and mercury, 0.0018.

Total average amounts of nitrogen added in the effluent and in the 25 kg/ha in the starter fertilizer annually were 105, 185, and 265 kg/ha for the 2.5, 5.0, and 7.5 cm/wk levels, respectively. The total annual average amount of inorganic nitrogen in kg/ha added from effluent, starter fertilizer, and precipitation in the 5 year period is presented in Table I.

Table I

Total Annual Inorganic Nitrogen Added (kg/ha) to Experimental Plots.

	Effluent Per Week (cm)		
Source	*2.5*	*5.0*	*7.5*
Effluent	80	160	240
Starter Fertilizer	25	25	25
Precipitation	22	22	22
Annual, Total	127	207	287

Objective 2.

The soil was sampled in the area described in objective 1 to a depth of 3.05 m in increments of 30.5 cm in 27 locations, 9 per replication, prior to establishment of the stands and annually thereafter at the end of the growing season. The sampling provided soil profile data on replicated treatments of effluent application level and on annual versus perennial species. Standard soil test information was obtained from 10 soil-depth increments at 27 locations to represent three levels of effluent addition with an annual and a perennial crop cover. Each of the three replications was sampled to add precision to the data. Soil data are reported for pH and extractable phosphorus.

Selected plant samples of the annual and perennial crops at the locations where soil profile samples were taken were analyzed for Kjeldahl nitrogen, phosphorus, aluminum, barium, calcium, magnesium, sodium, potassium, boron, copper, iron, manganese, and zinc by the appropriate micro-Kjeldahl, emission spectrographic, atomic absorption, and/or colorimetric analysis method.

Objective 3.

Porous vacuum-type samplers placed at depths of 15 and 150 cm were used to sample soil water before each weekly irrigation and 48 hr later. The sampling cups were placed in the center of large blocks of perennial grasses, perennial alfalfa, and annual grasses to prevent leachate from adjacent plots to affect the soil water.

DISCUSSION OF RESULTS AND THEIR SIGNIFICANCE

The highest yields were produced by the annual grasses such as corn, forage sorghum, and sorghum sudangrass. At the high level of effluent, these yields were 13.25 tonne/ha for corn, 15.38 tonne/ha for forage sorghum, and 12.56 tonne/ha for sorghum sudangrass. Yields of the annual grasses were obtained for only two of the five years because of severe damage to the corn by raccoons which caused excessive variations in yield.

Six alfalfa cultivars yielded twice as much as the best perennial grass (tall fescue) at the low (2.5 cm) and medium level (5.0 cm) of weekly effluent and nearly as much (10.31 tonne/ha) as tall fescue (11.03 tonne/ha) at the high level (7.5 cm) of effluent (Table II). The alfalfa cultivars were higher yielding than the grasses at the low and medium levels of effluent since these two levels had lower amounts of

Table II

Dry Matter Yields in Metric Tons Per Hectare of Forages Irrigated With Three Levels of Sewage Effluent in a 5 Year Period.

Species and Number of Varieties ( )	*Effluent Per Week (cm)**					
	2.5					
	1974	*1975*	*1976*	*1977*	*1978*	*5-Yr Ave.*
LEGUMES						
Alfalfa (6)	10.99	13.56	13.60	13.77	12.24	12.94
Birdsfoot Trefoil (2)	9.17	11.17	3.36	2.26	1.35	5.45
GRASSES						
Tall Fescue	8.16	8.72	6.61	6.70	6.12	7.26
Reed Canarygrass	7.38	8.92	4.87	5.83	4.69	6.32
Orchardgrass	7.73	8.05	5.25	5.74	4.33	6.21
Kentucky Bluegrass	4.53	5.02	2.76	5.04	3.77	4.21
Brome, Canadian	7.44	5.49	3.36	0.65	1.28	3.21
Reed Foxtail	4.53	5.18	3.30	2.71	1.01	3.14
Brome, Southern	7.15	6.46	0.90	0.70	0.72	3.18
Timothy	6.66	4.06	1.35	1.32	1.43	2.96

						2-Yr Ave.
ANNUALS						
Corn, Michigan 250 Total (grain, cob, stover)	15.42	--	--	11.25		13.34
Forage Sorghum	12.80	--	--	16.55		12.20
Sorghum Sudangrass	9.73	--	--	14.69		12.20
			5.0			
						5-Yr Ave.
LEGUMES						
Alfalfa (6)	11.77	13.90	13.47	10.72	10.38	11.66
Birdsfoot Trefoil (2)	11.23	12.67	3.14	2.13	1.46	6.13
GRASSES						
Tall Fescue	9.48	10.56	9.80	10.60	8.90	9.86
Reed Canarygrass	9.26	9.95	6.82	9.17	6.64	8.36
Orchardgrass	8.47	9.75	7.85	8.61	7.17	8.36
Kentucky Bluegrass	8.25	9.15	6.66	9.75	6.73	8.12
Brome, Canadian	10.18	4.15	1.04	0.79	0.79	3.39
Reed Foxtail	6.48	4.98	3.39	2.96	1.36	3.83
Brome, Southern	10.40	4.26	1.35	0.65	0.49	3.43
Timothy	6.57	4.48	1.30	1.39	0.67	2.88

						2-Yr Ave.
ANNUALS						
Corn, Michigan 250 Total (grain, cob, stover)	14.48	--	--	11.41		12.96
Forage Sorghum	11.91	--	--	14.37		13.14
Sorghum Sudangrass	10.76	--	--	14.26		12.51

7.5

						5-Yr Ave.
LEGUMES						
Alfalfa (6)	11.52	13.05	12.15	8.18	8.74	10.31
Birdsfoot Trefoil (2)	10.00	12.35	3.74	1.75	1.75	5.89
GRASSES						
Tall Fescue	9.44	13.05	10.83	11.97	9.84	11.03
Reed Canarygrass	7.76	12.82	9.48	11.19	8.70	9.98
Orchardgrass	8.74	10.87	9.98	10.76	8.27	9.78
Kentucky Bluegrass	5.54	9.35	8.21	10.65	7.96	8.59
Brome, Canadian	9.89	4.87	1.33	1.14	0.92	3.63
Reed Foxtail	5.67	5.13	3.72	2.02	1.35	3.57
Brome, Southern	9.64	4.55	0.95	0.76	0.40	3.26
Timothy	6.59	5.27	1.45	1.30	1.19	3.16

					2-Yr Ave.
ANNUALS					
Corn, Michigan 250 Total (grain, cob, stover)	12.73	--	--	13.77	13.25
Forage Sorghum	14.66	--	--	16.08	15.38
Sorghum Sudangrass	10.56	--	--	14.57	12.56

*1975 - July 17 to late October
1975-78 mid-May to late October

nitrogen which were inadequate for high yields of grass but not necessary for alfalfa which "fixed" its own nitrogen from the air when an adequate amount was not present in the effluent.

At the high level of effluent (7.5 cm/wk), which is the level most likely to be used in a wastewater renovation program, tall fescue yielded one tonne more than reed canarygrass, generally acclaimed to be the best species for wastewater renovation. Orchardgrass (9.78 tonne) had the same average annual yield as reed canarygrass (9.98 tonne) indicating its high yield potential. Since it is much more acceptable as an animal feed than reed canarygrass, its use in a wastewater renovation program may be important in the future. Kentucky bluegrass yielded 8.59 tonne, about one tonne less than orchardgrass. Its production was higher than expected because it is less vigorous and generally less productive than the other grasses which are much taller and productive.

In contrast to these high yields of tall fescue, reed canarygrass, orchardgrass and Kentucky bluegrass, the other four grasses (Canadian bromegrass, southern bromegrass, reed foxtail, and timothy) produced much lower yields in the magnitude of 3.0 to 3.5 tonne. These yields were considered too low for satisfactory crop yields. Undoubtedly these low yielding grasses would not have adequate top growth to remove enough nitrogen from the effluent (Table III). As a result, nitrate would accumulate excessively in the soil water.

Bromegrass, timothy, and reed foxtail began to thin out starting in the third year when irrigated at the high level. By the end of the fifth year, stands of these grasses were reduced to 10 percent or less of the original stand (Figure 1). In contrast, tall fescue, reed canarygrass, orchardgrass, and Kentucky bluegrass maintained excellent stands with 95 percent or more of the original stand at the end of the fifth year. All of these grasses had high yield levels indicating their potential for high production and excellent persistence under a system of wastewater management. The principal ingressing grass species was quackgrass *(Agropyron repens* L.*)*. Dandelion *(Taraxacum officinale* L.*)* was the primary broad-leaved ingressing species.

Alfalfa yields generally decreased as weekly effluent levels increased (Table IV, Figure 2). This indicates that alfalfa, adapted primarily to well-drained soil, will not tolerate high levels of effluent unless the cultivar is specifically developed to resist *Phytophthora megasperma*, (PRR) a root rot disease prevalent under high moisture conditions. Agate, developed specifically to resist PRR, was the highest yielding alfalfa cultivar for the 5 year period (Table IV). Equally importantly, it had the highest yield in the fifth year of all cultivars indicating its resistance to PRR. In the fifth year, Agate had an 87 percent stand compared to

Table III

Nitrogen Removed Annually in Forages of the High Yielding Legume, Perennial Grasses, and Annuals Under Three Effluent Rates For a 5 Year Period Compared to Total Nitrogen Added in Effluent, Starter Fertilizer, and Precipitation.

	Effluent Per Week (cm)		
Species	*2.5*	*5.0*	*7.5*
	----------*kg/ha*----------		
Agate Alfalfa	449	431	379
Reed Canarygrass	132	200	273
Tall Fescue	131	187	221
Orchardgrass	124	176	223
Kentucky Bluegrass	84	170	193
Corn, Michigan 250*	158	156	159
Forage Sorghum*	139	131	200
Sorghum Sudangrass*	115	125	163
Total Nitrogen Added, kg/ha	127	207	287

*2 Years, 1974 and 1977.

Vernal and Ramsey, the next most persistent cultivars, which had 67 percent of the original stand. Research in Michigan has shown that Hiphy, a cultivar also resistant to PRR, was 19 percent higher yielding than Agate in a 5 year test indicating that new cultivars are superior to Agate which was developed almost a decade ago. As in the perennial grasses, quackgrass and dandelion were the primary ingressing weedy species.

The two birdsfoot trefoil cultivars produced about one-half as much forage annually as alfalfa (5.89 *vs* 10.31 tonne/ha) in the 5 year period (Table II, Figure 2). Yields of trefoil cultivars were nearly equal to those of alfalfa in the first two years, but the stands of trefoil thinned severely starting in the third year (Figure 2). Stands of trefoil continued to deteriorate until the fifth year when the production was less than 2 tonne (Figure 2) compared to 8.72 tonne for the best cultivar of alfalfa (Agate, Table III). It was surprising that birdsfoot trefoil, notably resistant to wet land conditions, did not maintain its stand under any of the three levels of effluent. As in the perennial grasses and alfalfa, quackgrass and annual broad leaved weeds, primarily dandelion, invaded the trefoil stands.

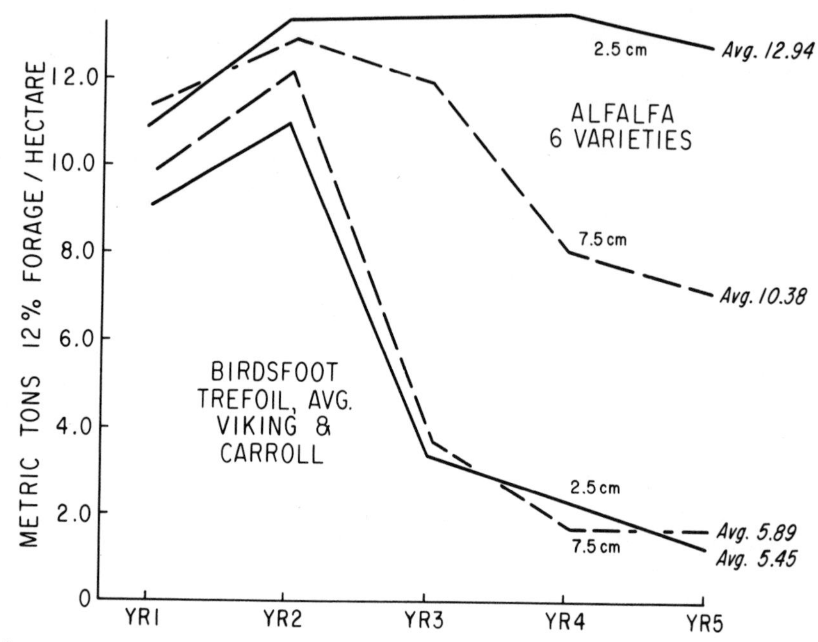

Figure 1. Production over a 5 year period of an average of six cultivars of alfalfa and two cultivars of birdsfoot trefoil when treated with 2.5 or 7.5 cm wastewater effluent per week from early May to October.

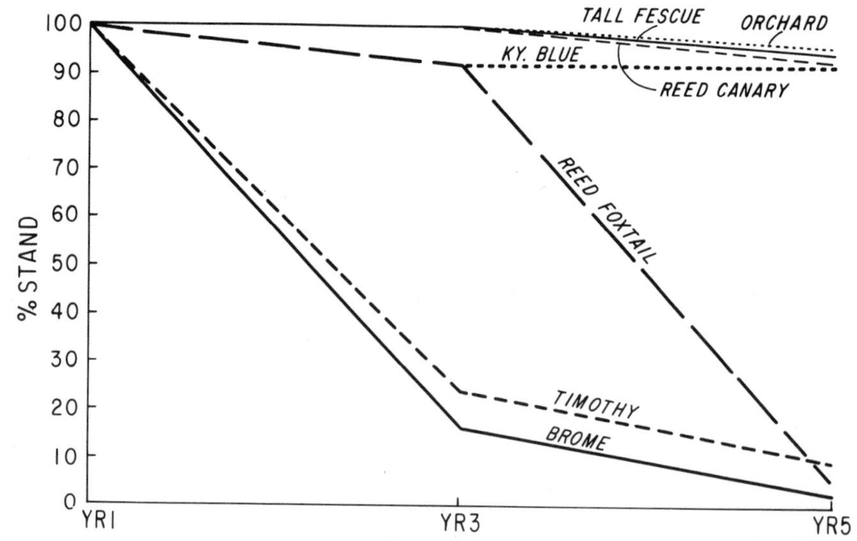

Figure 2. Percentage of original perennial grass species in sward over a 5 year period when treated with 7.5 wastewater effluent per week from mid may to October.

Nitrogen and phosphorus increased in percentage in orchardgrass and forage sorghum as rates of effluent increased during the 5 year period (Table V). Nitrogen did not increase in alfalfa as rates of effluent increased. This was expected since alfalfa is an effective nitrogen fixer. Phosphorus in alfalfa did increase with increasing levels of effluent in alfalfa, however. This was expected since three times as much phosphorus was applied in the high compared to the low level of effluent. Increases in effluent level were not reflected in any other major changes in percentage composition as indicated in Table V.

The concentration of mineral nitrogen in the soil water (Figures 3 to 8) verified the leaching which occurred with the annuals on week 21 of 1976 and week 18 of 1977. Mineral nitrogen, almost entirely nitrate-nitrogen, sharply increased in the root zone of the annuals (Figures 3 to 5). Nearly simultaneously, increases occurred 150 cm below surface of the 5.0 and 7.5 cm/wk applications (Figures 7 and 8). At the 2.5 cm/wk irrigation rate, only a gradual increase in mineral nitrogen concentration occurred at 150 cm (Figure 6). When irrigated at 2.5 cm/wk, the net recharge during May and June averaged 16.8 cm, while it averaged 39.2 and 56.3 cm for the 5.0 and 7.5 cm/wk rates, respectively. The low recharge prevented leaching of the excess mineral nitrogen from the topsoil to the groundwater. The inability of the annuals to use the excess nitrogen applied in these early growth months and the inability of the soil to retain nitrate against the leaching pressure of the higher application rates resulted in recharge of water containing nitrate in excess of 10 mg/l.

Once the annuals began to take up the nitrogen mineral nitrogen in soil water in the root zone (Figures 3 to 8) and subsequently at the 150 cm depth (Figures 6 to 8) diminished rapidly. The annuals were as effective as perennial grasses during this time in preventing nitrogen leaching. In fact, growth of corn on the 2.5 and 5.0 cm/wk plots was limited by availability of nitrogen. At those irrigation rates, yields of the corn were 15 and 14 percent lower than at the 7.5 cm/wk rate. Any nitrogen lost by leaching or denitrification would further widen the deficiency. Any decrease in yields due to nitrogen deficiency would be expected to lower the uptake of phosphorus and other nutrients as well.

With the perennial alfalfa, nitrate in the root zone increased to concentrations greater than 10 mg/l immediately following each harvest (Figures 3 to 5). As the alfalfa recovered, it again removed much of the added nitrogen and, overall, was effective in preventing excess nitrate leaching at all three irrigation rates (Figures 6 to 8). Because the alfalfa could fix nitrogen to make up any deficiency caused by low applications, yields were not affected by nitrogen levels. Rather the higher irrigation rates increased disease problems due to *Phytophthora megasperma* and lowered yields.

Table IV

Dry Matter Yields in Metric Tons Per Hectare of Six Cultivars of Alfalfa When Irrigated With Three Levels of Sewage Effluent in a 5 Year Period.

Alfalfa Variety	*Effluent Per Week (cm)*											
	1974				*1975*				*1976*			
	2.5	*5.0*	*7.5*	*Ave*	*2.5*	*5.0*	*7.5*	*Ave*	*2.5*	*5.0*	*7.5*	*Ave*
Agate	10.25	10.94	10.92	10.69	14.35	13.56	11.93	13.27	13.97	12.78	12.69	13.14
Iroquois	11.99	11.52	11.48	11.66	13.05	14.64	13.95	13.88	14.62	14.46	12.06	13.72
520	10.22	12.35	12.55	11.68	13.54	13.95	13.92	13.79	13.41	13.59	12.44	13.14
Vernal	11.46	12.96	11.68	12.04	13.61	13.79	12.78	13.38	14.01	13.16	12.51	13.23
Ramsey	11.10	10.90	11.12	11.03	12.02	12.98	11.97	12.33	12.26	13.74	11.95	12.64
Saranac	10.83	12.02	11.43	11.43	14.77	14.46	13.72	14.33	13.47	13.14	11.46	12.69
Average	10.99	11.77	11.52	11.43	13.56	13.90	13.05	13.52	13.61	13.47	12.15	13.07

	1977				*1978*				*5 yr. Ave., 74-79*			
Agate	12.33	10.54	9.21	10.69	12.51 (95)	10.56 (92)	8.72 (87)	10.58	12.69	11.66	10.69	11.68
Iroquois	15.11	11.84	8.54	11.83	13.50 (97)	9.71 (75)	5.92 (60)	9.73	13.50	12.33	10.38	12.07
520	14.48	10.81	7.94	11.08	12.94 (96)	8.30 (77)	4.66 (38)	8.68	13.36	11.88	10.27	11.83
Vernal	13.36	9.64	8.09	10.36	12.24 (90)*	10.38 (83)	6.19 (67)	9.60	12.94	12.11	10.22	11.76
Ramsey	13.68	11.55	8.09	11.10	11.59 (92)	10.02 (91)	5.09 (67)	8.90	12.31	12.11	9.80	11.40
Saranac	13.65	9.86	7.26	10.27	12.24 (98)	7.87 (63)	4.21 (50)	8.09	13.00	11.43	9.62	11.35
Average	13.77	10.72	8.18	10.90	12.50	9.47	5.80	9.26	12.96	11.92	10.16	11.68

*Percent legume in August 1978 in third cutting.

Table V

Chemical Composition, Dry Weight Basis, of the Third Cutting of Alfalfa and Orchardgrass (Mid October) and the Only Cutting of Forage Sorghum (Mid September).

Species and Variety	P	N	K	Na	Ca	Mg	Cu	Fe	Zn	B	Mn	Al	Ba
	%						mg/kg						
2.5 cm/wk													
Sorghum, Forage	0.19	0.9	1.80	0.04	0.48	0.18	8	305	18	10	17	100	9
Alfalfa, Saranac	0.15	2.6	1.35	0.26	0.46	0.26	8	146	12	56	18	98	11
Orchardgrass	0.33	2.4	2.01	0.14	0.76	0.25	7	251	16	14	30	101	12
5.1 cm/wk													
Sorghum, Forage	0.19	1.0	1.78	0.08	0.46	0.20	8	201	17	10	16	165	8
Alfalfa, Saranac	0.20	2.9	1.76	0.25	0.46	0.21	8	283	15	47	23	235	17
Orchardgrass	0.30	2.5	1.68	0.18	0.86	0.25	9	290	11	19	40	200	10
7.6 cm/wk													
Sorghum, Forage	0.20	1.3	1.60	0.08	0.39	0.17	9	300	19	8	22	200	7
Alfalfa, Saranac	0.17	2.9	1.80	0.27	0.57	0.25	8	152	11	61	22	125	11
Orchardgrass	0.29	2.7	1.59	0.26	0.90	0.33	11	200	16	20	40	140	10

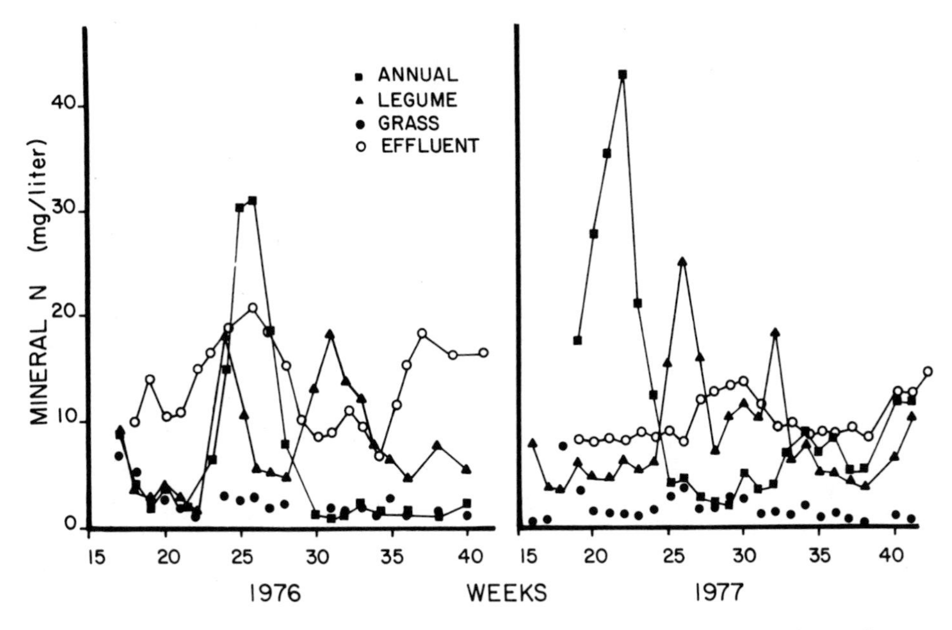

Figure 3. Weekly average mineral nitrogen concentrations in applied wastewater effluent and in soil water from the 15 cm depth of the 2.5 cm/wk irrigation rate of the various crop types.

The 1976-77 average annual yields were 12.96, 11.92, and 10.16 tonne/ha for the 2.5, 5.0, and 7.5 cm/wk irrigation rates, respectively (Table III).

Table III shows the average annual nitrogen removal by Agate (the best alfalfa cultivar), the four highest yielding perennial grasses, and the three annual grasses for the 5 year period in comparison to the total amount of nitrogen added from effluent, starter fertilizer, and precipitation. Agate alfalfa, a nitrogen fixing legume, had the highest amount of nitrogen in the plant tissue. It was not determined whether this was derived from the air or from nitrogen added as effluent, fertilizer, and in precipitation. Figures 6 to 8, however, show that the amount of nitrogen in the soil water at a depth of 150 cm after four years of forage production was below 10 mg/l at the high effluent rate of 7.5 cm/wk. This indicates that alfalfa was obtaining most of its nitrogen from the added nitrogen rather than from the air as it does under normal production if well inoculated with *Rhizobium meliloti*. Apparently the legume derived its nitrogen from applied nitrogen, the nitrogen source which expended the least energy in the plant. Alfalfa, therefore, was nearly as efficient as the perennial grasses in removing

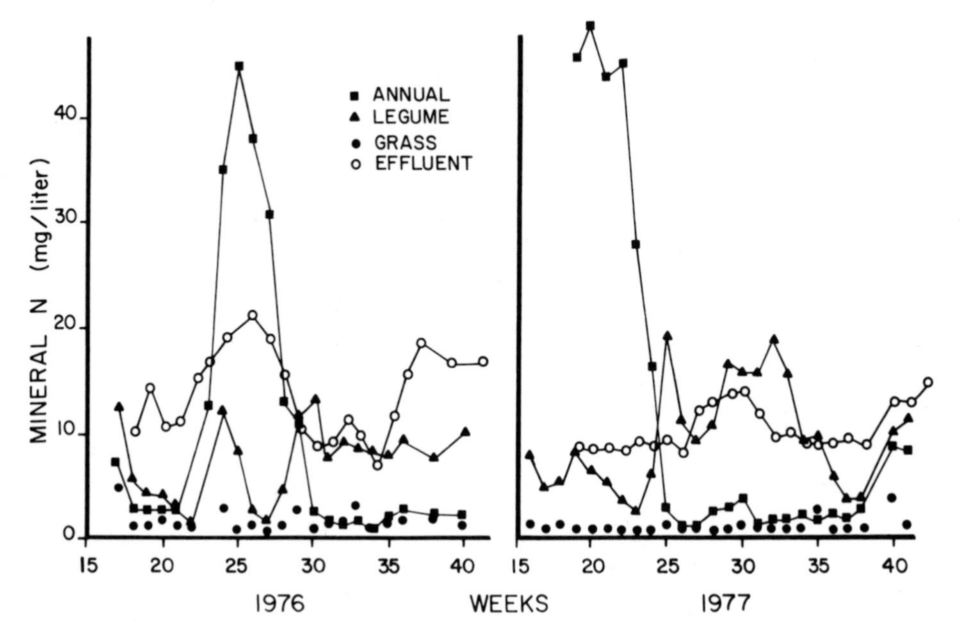

Figure 4. Weekly average mineral nitrogen concentrations in applied wastewater effluent and in soil water from the 15 cm depth of the 5.0 cm/wk irrigation rate of the various crop types.

nitrogen from applied nitrogen. The annual grasses, such as corn and the sorghums, were much less effective in removing nitrogen from the applied nitrogen than the legumes (Table III and Figure 8).

Most of the phosphorus in the effluent applied during the 5 year period and measured at the end of the experiment which was not removed by plants was concentrated in the upper 15 cm of the soil profile (Table VI). The amount varied from 29 at the low to 43 kg/ha at the high level of effluent compared to an initial base line of 12 kg/ha. This is in agreement with early work by Kardos and Sopper (1973) who showed that phosphorus accumulated in the upper 15 cm of the soil profile.

Table VI shows that as effluent levels increased, more nitrogen accumulated at lower levels: at the low, medium, and high levels of effluent, phosphorus accumulated in the profile to depths of 61, 91, and 122 cm, respectively. The amounts in these levels between 15 and 122 cm, however, were only slightly higher than the base line data at the start of the experiment. Table VI shows there was no accumulation of phosphorus below the level of 122 cm even at the highest effluent level.

Table VI

Average Phosphorus for All Legumes, Perennial Grasses, and Annual Grass Crops as Measured at the Beginning and End of the Experiment in Which Three Levels of Effluent Were Applied Annually in a 5 Year Period.

Soil Increment Depth	*1973 Base Line*	*1978 Effluent Per Week (cm)*		
		2.5	*5.0*	*7.5*
---*cm*---	------------------*kg/ha*------------------			
0- 15	12	29	44	43
15- 30	2	9	7	8
30- 61	1	6	6	6
61- 91	1	2	5	6
91-122	1	2	2	6
122-161	1	2	2	3
162-183	1	2	2	3
183-213	1	2	2	2
213-244	1	2	2	2
244-274	1	2	2	2
274-305	1	2	2	2

Reed canarygrass had the greatest amount of nitrogen in the plant tissue (Table III) of all perennial and annual grasses at all three effluent rates. The nitrogen removal of all four of the perennial grasses, however, was high, ranging between 193 and 273 kg/ha at the high effluent rate. The perennial grasses had greater total nitrogen removal than the annual grass crops (*e.g.*, corn) as measured by nitrogen in the plant tissue at the two highest effluent rates. Reed canarygrass removed approximately as much nitrogen as was added to the high level of effluent. Nitrogen in the soil was well below the 10 mg/l standard at all three levels of effluent.

CONCLUSIONS

Corn, alfalfa, and orchardgrass are three valuable feed crops which produce high yields of high quality feed for livestock and can effectively remove phosphorus and nitrogen from municipal sewage water applied to a well-drained soil with a water table at 3 m.

Seven and one half cm was the high level of wastewater effluent applied each week between early May to late October

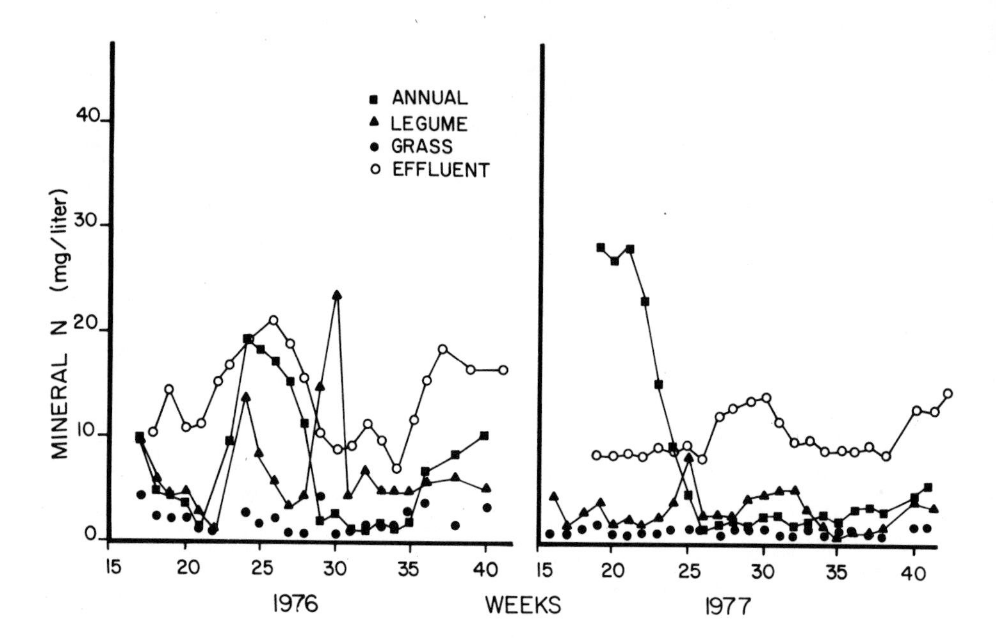

Figure 5. Weekly average mineral nitrogen concentration in applied wastewater effluent and in soil water from the 15 cm depth of the 7.5 cm/wk irrigation rate of the various crop types.

during five years of tests between 1974 and 1979 at Michigan State University. The phosphorus in the sewage effluent from East Lansing was used by plants for growth and was tied up primarily in the surface 15 cm of soil and did not contaminate groundwater or streams. Likewise, nitrogen was used by the grass crops, such as corn and orchardgrass, and the legume alfalfa to produce yields above the state average. The nitrogen which was not used by the crops and went below the root zone was well below the level of 10 mg/l designated by the U.S. Environmental Protection Agency (USEPA) as being safe for drinking water.

The suprise in renovating wastewater was alfalfa which, next to corn, is the primary feed for dairy cattle in the United States. Alfalfa normally takes "free" nitrogen from the air and fixes it in the plant. It is known as a "nitrogen fixer". In these experiments on a well-drained soil, alfalfa yielded more than perennial grasses but less than corn. It apparently did not fix much free nitrogen, getting most of its nitrogen from the sewage effluent. Agate, the best alfalfa cultivar, was resistant to root rot *(Phytophthora megasperma)* and produced twice as much in the fifth

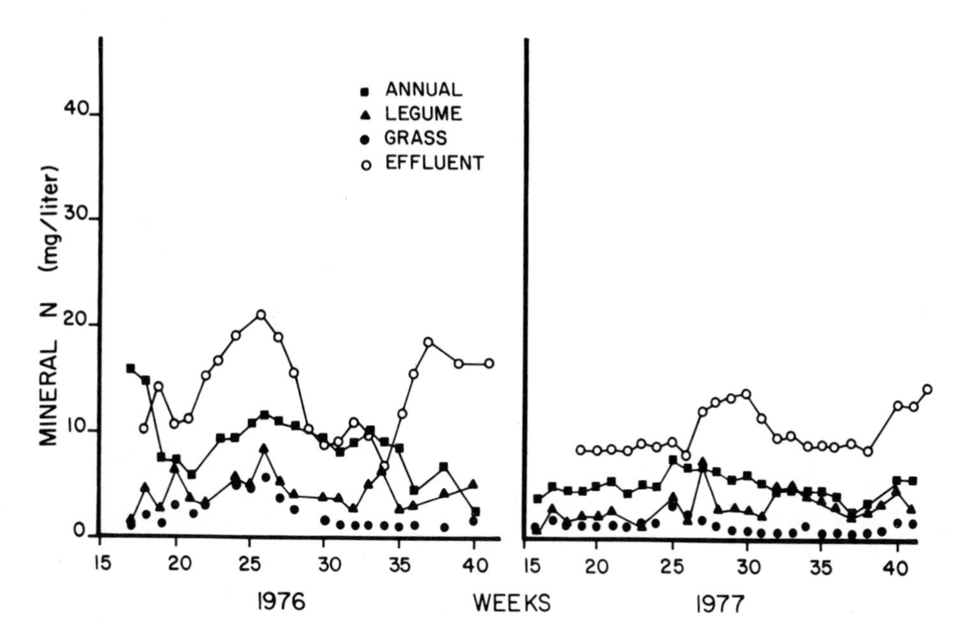

Figure 6. Weekly average mineral nitrogen concentrations in applied wastewater effluent and in soil water from the 150 cm depth of the 2.5 cm/wk irrigation rate of the various crop types.

year as the average of alfalfa produced in the Great Lakes area, indicating its perennial nature under high water on a well drained soil if resistant to root rot.

Orchardgrass was another surprise. It survived as well as reed canarygrass, commonly known to have high resistance to wet soils and yielded as well as canarygrass. Because of its greater animal acceptance, it is much more valuable in livestock production than reed canarygrass in the United States or tall fescue in the north central and northeastern states.

Municipalities can effectively and economically dispose of their wastewater by applying it to land owned by the municipality or piping it to farmers who would apply the wastewater to their crops and possibly pay for the wastewater. The phosphorus and nitrogen in the wastewater would be utilized by the plant or remain in the soil at a level below USEPA accepted standards. In areas of low rainfall, water in the wastewater would be even more valuable than humid areas such as Michigan. Corn is the best crop to remove nitrates in July and August because of its rapid growth in the hot summer. Perennial alfalfa and orchardgrass could be used in the spring and fall, before and after corn's peak need, to

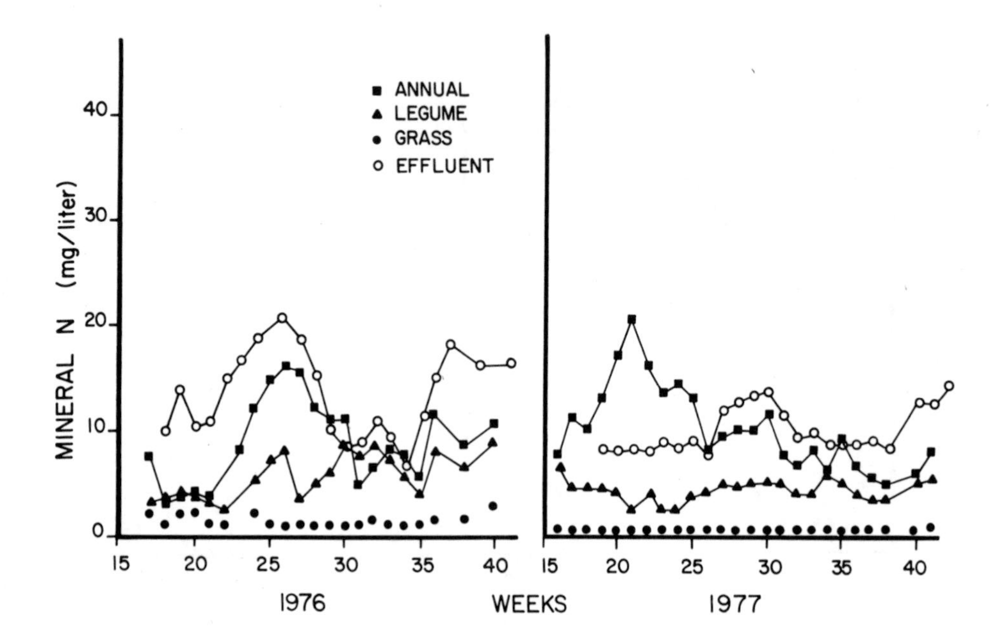

Figure 7. Weekly average mineral nitrogen concentrations in applied wastewater effluent and in soil water from the 150 cm depth of the 5.0 cm/wk irrigation rate of the various crop types.

remove the nitrates and phosphates from the wastewater. A combination of corn, alfalfa, and/or orchardgrass, all valuable feed crops for livestock, would be irrigated with municipal sewage effluent from early April to late October. All three crops, corn, alfalfa, and to a lesser extent, orchardgrass, are widely used by livestock farmers in their area of adaptation.

RECOMMENDATIONS

The following are specific recommendations, primarily for the north central and northeastern states, of valuable, widely used feed crops which are recommended for removing phosphorus and nitrogen from wastewater applied at a rate of 7.5 cm/wk from spring until late October when perennials become dormant in the northern half of the eastern United States:

1. Corn and alfalfa are first choice in their area of adaptation on well-drained soil (water table at 3 m). These two crops provide most of the feed for livestock in the north central and northeastern

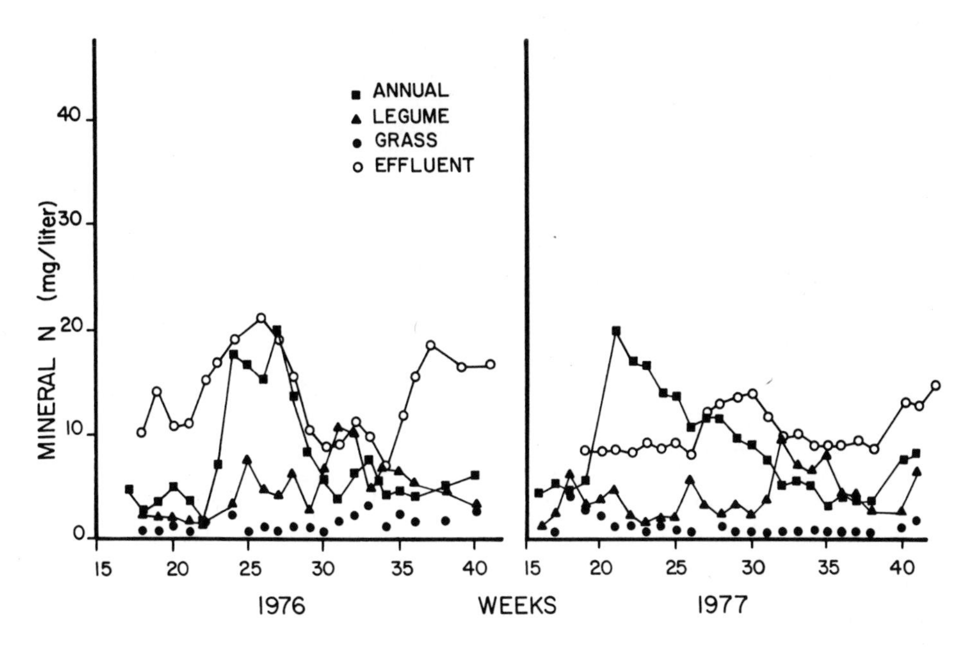

Figure 8. Weekly average mineral nitrogen concentrations in applied wastewater effluent and in soil water from the 150 cm depth of the 7.5 cm/wk irrigation rate of the various crop types.

states. They are highly digestible and are the primary feed crops for dairy milking herds in these areas. Corn will grow satisfactorily on artificially tiled soil where the water table is at a depth of 1.5 to 2 m but alfalfa, a perennial legume, requires better drainage for persistence. A water level at a depth of 3 m on untiled land provides such a drainage condition.

2. Under ideal conditions of soil where the water table on naturally well drained soil is at a depth of 3 m, alfalfa should persist for 5 years, possibly more, when treated with 7.5 cm wastewater per week during the growing period. A recommended time of starting irrigation would be in early May continuing through October when alfalfa becomes dormant. Alfalfa could be used for hay, silage, or pasture (poorer persistence than when used for hay or silage) for any class of livestock.

    The alfalfa cultivar selected should be high yielding with resistance to root rot *(Phytophthora megasperma)* and bacterial wilt *(Corynebacterium insidiosum* McCull H.L. Jens*)* in the alfalfa growing region. Resistance to anthracnose *(Colletotrichum trifolii* Bain*)* in all areas except the northern tier of states in this area would improve stand persistence and yield. Cultivars available today are more productive than, and equally as resistant to PRR, as Agate used in these trials.

3. Orchardgrass is first choice of the perennial grasses for renovating wastewater. It removed nitrogen from the wastewater nearly as well as canarygrass but is much more widely used and accepted as an animal feed than is reed canarygrass which is not popular with livestock farmers. Orchardgrass could be grown alone or in combination with alfalfa. Irrigation with wastewater could begin in April on orchardgrass which is more likely than alfalfa to tolerate high effluent levels.
4. A mixture of alfalfa and orchardgrass is recommended on soils which are not naturally well drained. The alfalfa may not persist over two years with a 7.5 cm/wk application, but orchardgrass would fill in the areas in the thinned alfalfa stand.
5. Tall fescue would be second choice after orchardgrass. It is preferred over reed canarygrass because it has greater acceptance as a livestock feed. It would not be acceptable in the northern tier of states because of its occasional lack of winter hardiness.
6. Kentucky bluegrass is more limited in use than the other grasses because of its lower yield and lower removal of nitrogen by the forage. It is of excellent quality for grazing, however, and would fulfill many requirements of a good grass for grazing by livestock such as horses.
7. The quality of the grasses sward could be increased by an annual broadcast seeding of red clover *(Trifolium pratense)* or Ladino clover *(Trifolium repens)* on the surface of the sward. Timely cutting or grazing to reduce grass competition under a system of grazing or mechanical harvesting would be necessary for good legume establishment.
8. If forage utilization and value are not important in selection of a perennial grass, reed canarygrass is first choice in its area of adaptation because of high yields, excellent removal of nitrogen from wastewater, winter hardiness, and persistence. It would produce a high yield suitable for animal feed of lower quality than orchardgrass or possibly for alcohol production.

Tall fescue, generally considered to be less acceptable than orchardgrass but more acceptable than reed canarygrass for animal feed, may be a good choice in the southern and eastern United States where orchardgrass becomes heavily infested with leaf diseases and is not productive.

9. Forage sorghum and sorghum sudangrass should be considered as alternatives to corn since they were tolerant of high rates of wastewater and produced yields of dry matter equal to or better than corn. They are second choice as annuals to corn since the dry matter produced by corn is about 60 percent corn grain which is more valuable than forage material. The sorghums produced very little grain in this experiment but are better grain producers in states with longer growing seasons where they are better adapted.
10. Corn (C) and alfalfa (A) grown alone, or orchardgrass (G) grown alone or with alfalfa could be used as indicated below to produce high quality animal feed in a system of wastewater application of 7.5 cm/wk on a well drained soil having a water table at a depth of 3 m. Corn, an annual, and orchardgrass, which is more tolerant than alfalfa of wet soil, would grow well with a water table at 1.5 to 2 m.

	*North Central Area**		*Northeastern Area**
Month	*Northern Half*	*Southern Half*	
March	--	G	--
April	G	G or A	G or A
May	G or A	A or G	G or A
June	A or G	A or G	A or G
July	C, A, or G	C, A, or G	C, A, or G
August	C, A, or G	C, A, or G	C, A, or G
September	C, A, or G	C, A, or G	C, A, or G
October	G or A	G or A	G or A
November	--	G	--

*Reed canarygrass or tall fescue could be grown instead of orchardgrass where adapted in these areas for lower quality feed or as a substrate for a alcohol production.

LITERATURE CITED AND REFERENCES

Brown, B.A. and R.L. Munsell. 1963. Penetration of Surface Applied Lime and Phosphate in the Soil of Permanent Pastures. Storrs Agr. Exp. Sta. Bull. 186.

Clapp, C.E., D.R. Linden, W.E. Larson, and G.C. Marten. 1977. Nitrogen Removal From Municipal Wastewater Effluent by a Crop Irrigation System. In R. Loehr (ed.), Land as a Waste Management Alternative. Ann Arbor Science Publishers, Inc., Ann Arbor, MI 48106, pp. 139-150.

Finn, B.M., S.J. Bourget, K.F. Nielsen, and B.K. Dow. 1961. Effects of Differential Soil Moisture Tensions on Grass and Legume Species. Can. J. Soil Sci. 41:16-23.

Gardner, W.R. 1965. Movement of Nitrogen in Soil. In Soil Nitrogen. Agronomy Monograph No. 10. American Society Agronomy Press, Madison, WI 53706.

Hook, J.E. and L.T. Kardos. 1977. Nitrate Relationships in Penn. State "Living Filter" System. In R. Loehr (ed.), Land as a Waste Management Alternative. Ann Arbor Science Publishers, Inc., Ann Arbor, MI 48106, pp. 181-197.

Ingalls, J.R., J.W. Thomas, E.J. Benne, and M.B. Tesar. 1965. Comparative Response of Wether Lambs to Several Cuttings of Alfalfa, Birdsfoot Trefoil, Bromegrass, and Reed Canarygrass. J. Anim. Sci. 24:1159-1163.

Kardos, L.T. and W.E. Sopper. 1973. Renovation of Municipal Wastewater Through Land Disposal by Spray Irrigation. In W.E. Sopper and L.T. Kardos (eds.), Recycling Treated Municipal Wastewater and Sludge Through Forest and Cropland. Pennsylvania State University, University Park, PA 16802, pp. 148-163.

Marten, G.C., C.E. Clapp, and W.E. Larson. 1979. Effects of Municipal Wastewater Effluent and Cutting Management on Persistence and Yield of Light Perennial Forages. Agron. J. 71:650-658.

Miller, R.H. 1977. The Soil as a Biological Filter. In R. Loehr (ed.), Land as a Waste Management Alternative. Ann Arbor Science Publishers, Inc., Ann Arbor, MI 48106, pp. 70-94.

McKenzie, R.E. 1951. Ability of Forage Plants to Survive Early Spring Flooding. Sci. Agr. 31:358-367.

Melbourne Board of Works Farm. 1969. Waste Into Wealth. Melbourne and Metropolitan Board of Works unnumbered report, Melbourne, Australia.

Neller, J.R. 1947. Mobility of Phosphates in Sandy Soils. Soil Sci. Soc. Amer. Proc. 11:227-230.

O'Donovan, P.B. 1964. *Ad libitum* Intake and Digestibility of Forages by Lambs as Related to Soluble and Structural Components. Ph.D. Thesis, Purdue University, Lafayette, IN 47907.

Powell, R.D. and L.T. Kardos. 1969. Effect of Moisture Regimes and Harvests on Efficiency of Water Use by Ten Forage Crops. Soil Sci. Soc. Amer. Proc. 32:871-874.

Tesar, M.B. 1976. Alfalfa Management for Maximum Returns in Michigan. Report of the 25th Alfalfa Improvement Conference, July 13-15, Ithaca, NY 14853.

Tesar, M.B. 1976. Yield and Nutrient Removal of Alfalfa Grown at High Yield Levels. Mich. Agr. Expt. Sta. Res. Rept. 176:25-34.

Tesar, M.B. 1976. Productivity of Birdsfoot Trefoil in Michigan. Mich. Agr. Exp. Sta. Res. Rept. 176:35-39.

Tesar, M.B. 1976. Fertilizer Nitrogen Versus Nitrogen From Legumes. Proc. 22nd Annual Farm Seed Conference, November 16, Kansas City, MO, Publication No. 22.

Tesar, M.B. 1978. Recommended Alfalfa Varieties for Michigan. Mich. Coop. Ext. Serv. Bull. E-1017.

Tesar, M.B. 1978. Sod Seeding Birdsfoot Trefoil and Alfalfa. Mich. Coop. Ext. Serv. Bull. 956.

Tesar, M.B. 1980. Clear Seeding of Alfalfa. Mich. Coop. Ext. Ser. Bull. E-961.

Thomas, J.W., A.D.L. Gorrill, G.N. Blank, and M.B. Tesar. 1965. Acceptability of Alfalfa, Bromegrass, and Canarygrass by Sheep and Their Performance. J. Anim. Sci. 24: 911.

CHAPTER 7

OLDFIELD MANAGEMENT STUDIES ON THE WATER QUALITY MANAGEMENT FACILITY AT MICHIGAN STATE UNIVERSITY

Thomas M. Burton, Department of Zoology, Department of Fisheries and Wildlife, and Institute of Water Research, Michigan State University, East Lansing, Michigan 48824; and James E. Hook, Department of Agronomy, Coastal Plains Experiment Station, Tifton, Georgia 31794

INTRODUCTION

The use of abandoned farm fields (oldfields) for tertiary treatment of secondary municipal wastewater was investigated on the Water Quality Management Facility (WQMF) at Michigan State University from 1975 through 1980. The purpose of these studies was to determine if spray irrigation on oldfields was a reasonable means of achieving tertiary treatment; and, if so, what the best plan of operation would be to achieve maximal renovation per unit area of oldfield. There were three aspects to these studies. First, a mass balance approach was used to investigate removal of nitrogen and phosphorus by the oldfield under various levels of irrigation and harvest or mowing during the growing season. Second, changes in the plant community resulting from these various treatments was monitored in detail. Third, feasibility of winter irrigation was investigated. The results of the mass balance studies have been reported by Burton and Hook (1978a), Hook and Burton (1978, 1979), Burton (1979), and Burton and King (1980). The changes in the vegetation have not been reported previously except in a final completion report with limited availability (Burton, 1978). Preliminary studies on the feasibility of winter irrigation was published by Leland *et al*. (1979). The purpose of this paper is to summarize these studies using data from the previous publications plus recent, unpublished data.

DESCRIPTION OF STUDY AREA

The WQMF was a mixture of abandoned farm fields and woodlots when wastewater research began on the site in 1974. The site is located on the southern edge of the Michigan State University campus about 8 kilometers south of East Lansing, Michigan. The abandoned fields had not been farmed for approximately 10 to 15 years and were dominated by a mixture of quackgrass *(Agropyron repens)* and goldenrod *(Solidago graminifolia* and *Solidago canadensis)* with a diverse mixture of other oldfield species. Some shrubs had started to invade these fields.

Soils of the site are members of the fine loamy mixed mesic family of Typic Hapludalfs and are highly heterogeneous. In general, they are well drained. Infiltration rates vary from 0 to over 20 cm/wk but average about 5 cm/wk over the entire WQMF. Infiltration rates on the oldfield plots were slightly greater than 10 cm/wk, the maximum rate of application of wastewater possible on these fields. Three study areas will be included in this paper. The majority of research has been conducted on 0.07 ha plots in one oldfield area (the oldfield plot studies). Studies on other sites included a small watershed site which was underlain by impermeable clays with infiltration rates of less than 5 cm/wk so that most losses occurred by runoff (the oldfield watershed study) and a small watershed used to investigate the feasibility of winter irrigation (the winter spray site). All of these sites had still functional tile drains draining portions of the site. Runoff from these drains was monitored for the oldfield plot studies and for the oldfield watershed. However, the drains were disrupted on the winter spray site and a small surface channel was dug to intercept water which ponded in the low area of this watershed.

Climate for the area alternates between continental and semimarine due to the influence of the Great Lakes. Precipitation averages 77 cm/yr with average snowfall being about 124 cm/yr. Average growing season extends from May 7 to October 8 (154 days), and annual temperature averages 8.2°C with a mean monthly low of -5.5°C in January and a mean monthly high of 21.6°C in July.

METHODS

Methods have been reported in detail elsewhere (Burton, 1978; Hook and Burton, 1979; and Leland *et al.*, 1979). Briefly, the effects of irrigation rate and harvest or mowing was investigated using 24 x 27 m plots in a randomized block sampling design with irrigation rates of no irrigation, 5 cm/wk, and 10 cm/wk of irrigation applied in two applications per week with a surface agricultural type spray irrigation system at rates of 0.83 cm/hr. For each irrigation rate, there were four replicates each of no harvest, one harvest in June, or two harvests in June and September for a total of 36 plots. Leaching from each of these plots was monitored with porous cup vacuum-type soil water samplers placed at depths of 15 and 120 cm. Inputs of nitrogen, phosphorus, and chloride in wastewater were monitored by collecting samples with funnels in the field about 1 m above the vegetation. Runoff was monitored using hydrologic recorders and ISCO sequential samplers. Mass balances were constructed by monitoring the quantity and quality of inputs in rain and wastewater, by calculation of evapotranspiration using the technique of Thornthwaite and Mather (1967), and by assigning the difference in quantity to groundwater seepage. This quantity of groundwater times average concentration of nitrogen, phosphorus, and chloride in the 120 cm depth soil water samples allowed estimates of leaching rates. Removal of nitrogen and phosphorus in vegetation was also measured and included in the mass balances. Runoff was less than 10 percent of annual losses of water; and since the contributing areas could not be determined, all losses were assigned to groundwater seepage. Estimates of runoff losses will be included in this paper for the entire field.

Vegetation changes were determined by taking four random 0.25 m^2 quadrat samples per plot every 4 to 6 weeks throughout the growing season, sorting to species, drying, grinding in a Wiley mill, and subsampling for chemical analyses (Burton, 1978). Also, samples of harvested vegetation were taken for chemical analysis.

In addition to the oldfield plot studies described below, the winter irrigation study was conducted using similar techniques for a 3 ha watershed (Leland *et al.*, 1979) as was oldfield studies on a separate 7.7 ha small watershed. Only 5.3 ha of this latter site was irrigated. These three areas will be described as the oldfield plot studies, the oldfield winter spray site, and the oldfield watershed for the rest of this paper. Pre-irrigation studies of runoff from the oldfield watershed have been reported by Burton and Hook (1978b, 1979).

RESULTS

Mass Balance Studies

The primary concerns for land application systems are (1) prevention of losses of nitrate in excess of the drinking water standard of 10 mg N/1 to either the groundwater or in runoff, (2) prevention of excessive runoff of phosphorus (in Michigan, the standard is 1 mg P/l), (3) prevention of losses or excessive accumulation in soils and plants of toxic substances such as heavy metals and toxic organics, and (4) prevention of pathogenic organism movement in leachate, runoff, or aerosol losses. Since wastewater from East Lansing contains little heavy metal or organic toxicant pollution, the research emphasis on the WQMF has been on nitrogen and phosphorus. No major study of the human pathogens has been made, although there are a few preliminary data. Studies on plant pathogens are included in this volume by Epstein and Safir (see Chapter 12).

Mass balances constructed for the oldfield plot studies demonstrated that oldfields are capable of preventing excessive losses of inorganic nitrogen (Table I). In fact, concentrations in leachate remained well below 10 mg N/l throughout most of the growing season even at the highest rate of irrigation (Figure 1). Harvesting twice removed most of the applied nitrogen even at the 10 cm/wk irrigation rate (Table I). Harvesting only once was less effective. Even on the unharvested plots in 1976-77, most of the applied nitrogen was retained on-site (Table I). Most of the nitrogen losses on these plots occurred after growth of vegetation ceased in mid-August, from mid-August through October (Figure 1). The effect of harvest appeared to be to prolong the active growing season resulting in plant uptake and on-site retention well into October (Figure 1). These data suggested that mowing might be as effective as harvest in preventing nitrogen losses. This alternative was investigated from 1978 through 1980. The 1978 data were encouraging (Figure 1), although there was a trend towards increased losses on the one mowing treatment for the 10 cm/wk application rate (Table I).

One disturbing trend was the tendency towards increased leaching losses on the unharvested plots from 19 percent of input with harvest to 45 percent of input with mowing only for the 5 cm/wk treatment from the 1977 to 1978 growing seasons and from 44 to 61 percent of input for the 10 cm/wk treatment (Table I). This trend suggests that the capacity of the site to retain nitrogen may be limited and that losses will continue upwards until inputs eventually equal outputs. However, concentrations have remained low through the 1980 growing season with annual averages of 1.93 ± 1.73 and 1.81 ± 1.16 mg N/l for the 5 and 10 cm/wk irrigation rates, respectively. Thus, these plots still meet standards without harvest or mowing after 6 years of spray irrigation.

On-site retention has to be limited unless significant losses are occurring by denitrification. We have not studied this possibility but the losses of 45 and 61 percent of input on the unmowed plots in 1978 suggest that it is not great enough, if it occurs, to prevent substantial leaching. Also,

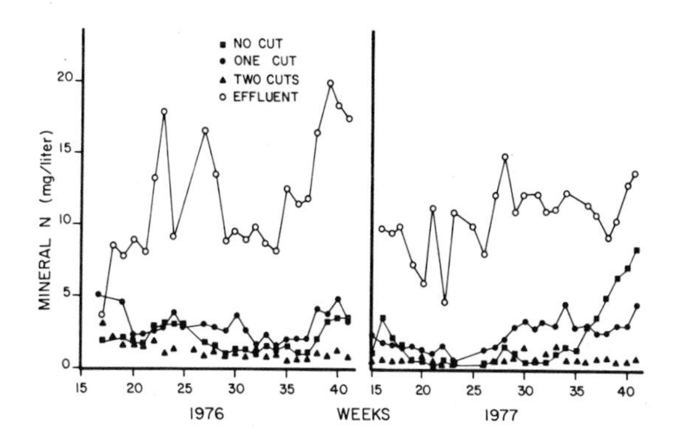

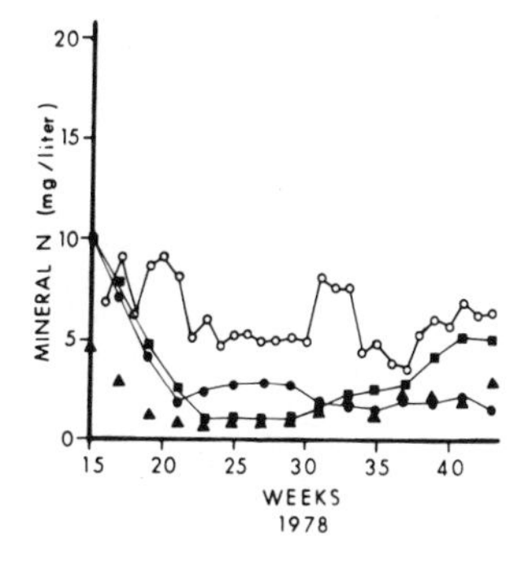

Figure 1. Average weekly inorganic nitrogen concentrations in applied wastewater and in soil water from the 120 cm depth, 10 cm/wk irrigation rate oldfield site. From Burton (1979) with the 1976 and 1977 data modified from Hook and Burton (1979).

the excellent retention in 1978 on sites where there was negative on-site retention in 1977 due to harvest and leaching coupled with the tendency towards increased losses on the unharvested plots suggests that on-site retention is linked to storage in live and dead organic matter and that there is a finite capability for such storage. Thus, breakthrough of nitrate at levels near input levels could occur in the future unless the sites are harvested.

About 99 percent of applied phosphorus was retained on-site for the unharvested plots (Table II). This retention was probably by sorption on soil as one would expect. The effect of harvest was to remove significant quantities of phosphorus in harvested biomass, thus prolonging the sorption capacity of the site (Table II). Two harvests removed 73 and 41 percent of applied phosphorus for the 5 and 10 cm/wk sites, respectively (Table II). Concentrations of soluble reactive

Table I

Mass Balances for Inorganic Nitrogen for the 5 and 10 cm/week Oldfield Plot Studies for the 1976-77 and 1977-78 Water Years, the Third and Fourth Years of Effluent Irrigation (Modified From Hook and Burton, 1979; Burton, 1979).

	*Total Inputs**	*Harvest Removal*	*Leaching*	*On-Site Retention*
	*5 cm/wk - No Harvest/Mowing Treatment***			
1976-77 (kg/ha/yr)	151.3	0	28.8	122.5
% of input	100.0	0	19.0	81.0
1977-78 (kg/ha/yr)	115.0	0	52.2	62.8
% of input	100.0	0	45.4	54.6
	*5 cm/wk - One Harvest/Mowing Treatment***			
1976-77 (kg/ha/yr)	151.3	128.0	30.2	-6.9
% of input	100.0	84.6	20.0	-4.6
1977-78 (kg/ha/yr)	115.0	0	25.3	89.7
% of input	100.0	0	22.0	78.0
	*5 cm/wk - Two Harvest/Mowing Treatment***			
1976-77 (kg/ha/yr)	151.3	156.0	10.2	-14.9
% of input	100.0	103.1	6.7	-9.8

1977-78	(kg/ha/yr)	115.0	0	5.3	109.7
	% of input	100.0	0	4.6	95.4
*10 cm/wk - No Harvest/Mowing Treatment***					
1976-77	(kg/ha/yr)	268.4	0	119.2	149.2
	% of input	100.0	0	44.4	55.6
1977-78	(kg/ha/yr)	203.1	0	123.7	79.4
	% of input	100.0	0	60.9	39.1
*10 cm/wk - One Harvest/Mowing Treatment***					
1976-77	(kg/ha/yr)	268.4	147.0	75.8	45.6
	% of input	100.0	54.8	28.2	17.0
1977-78	(kg/ha/yr)	203.1	0	114.6	88.5
	% of input	100.0	0	56.4	43.6
*10 cm/wk - Two Harvest/Mowing Treatment***					
1976-77	(kg/ha/yr)	268.4	237.0	87.1	-55.7
	% of input	100.0	88.3	32.5	-20.8
1977-78	(kg/ha/yr)	203.1	0	59.9	143.2
	% of input	100.0	0	29.5	70.5

*Includes inputs of 5.3 and 5.0 kg/ha/yr in precipitation for the 1976-77 and 1977-78 water years, respectively.

**The plots were harvested during the growing seasons of 1976 and 1977 but were mowed with no vegetation removal during the 1978, 1979, and 1980 growing seasons.

phosphorus remained low in soil water at the 120 cm depth with annual mean concentrations varying from 0.20 ± 0.01 for the unirrigated, unharvested area to highs of 0.036 ± 0.11 for the 5 cm/wk irrigation with one harvest treatment. No significant trends in concentration versus treatment were apparent (Hook and Burton, 1978).

Table II

Phosphorus Mass Balance for the 1975-76 Water Year for the Various Treatments of the Oldfield Plot Studies (Values in kg P/ha) (Hook and Burton, 1978).

Irrigation Rate	*Harvests*	*Effluent Input*	*Harvest Removal*	*Leaching*	*On-Site Retention*
0 cm/wk	None	0.0	0.0	0.01	0.0
	One	0.0	9.5	0.01	-9.5
	Two	0.0	15.6	0.01	-15.6
5 cm/wk	None	33.3	0.0	0.40	32.9
	One	33.3	14.4	0.45	18.5
	Two	33.3	24.2	0.19	8.9
10 cm/wk	None	65.8	0.0	0.55	65.3
	One	65.8	19.2	0.67	45.9
	Two	65.8	26.9	0.51	38.4

Runoff could not be apportioned on a per plot basis for the oldfield plot studies because of the unknown drainage field of the existing tile system. However, if one considers application to the entire 3.4 ha irrigated area at the rate of 7.5 cm/wk (5 for half the area, 10 for the other half) for a 26 week period, inputs of water would be about 19,500 m^3/ha/yr. Runoff was only 2,350 m^3/ha for 1978-1979 (Table III) or 12.1 percent of input. Likewise, if one considers inputs of inorganic nitrogen in irrigation and precipitation of about 150 kg N/ha/yr (Table I), export of inorganic nitrogen of 7.6 kg/ha/yr represents about 5.1 percent of input. On the same basis, annual export of phosphorus is about 4.9 percent of input and about 12.6 percent of applied chloride is lost in runoff. Thus, assigning all water not lost by evaportranspiration to seepage on the mass balance calculations for individual plots should not significantly affect these mass balances. Runoff losses for various constituents for the study sites are summarized in Tables III and IV.

Table III

Export of Nitrogen, Phosphorus, and Chloride in Runoff From Three Oldfield Sites From October 1, 1978, to September 30, 1979. Values Are kg/ha Unless Otherwise Indicated.

	Oldfield Plots	*Oldfield Watershed*	*Winter Spray Oldfield*
Inorganic N	7.59 ± 3.71	1.47 ± 0.09	8.47 ± 2.71
Ammonia N	0.54 ± 0.13	0.66 ± 0.04	1.47 ± 0.91
Nitrate N	6.66 ± 0.96	0.59 ± 0.09	6.67 ± 2.31
Nitrite N	0.39 ± 0.11	0.21 ± 0.03	0.46 ± 0.20
Organic N	3.21 ± 0.45	9.96 ± 0.29	7.07 ± 1.87
Total P	2.45 ± 0.35	1.78 ± 0.10	3.81 ± 1.19
Soluble Reactive P	1.93 ± 0.32	0.83 ± 0.06	2.99 ± 0.09
Chloride	219.29±10.92	693.64±11.41	767.06±111.98
Discharge (m^3/ha)	2350	9580	7838

Concentration in runoff does sometimes exceed standards for phosphorus during peak runoff as the existing tiles appear to intercept a small amount of the wastewater during application with little soil contact time, perhaps because of subsurface channels or because of sand lenses (Figure 2). Several holes leading directly to the tiles were discovered and filled during the course of the study. These high concentrations during peak discharge result in runoff exceeding standards for phosphorus about 11 percent of the time for the oldfield plot site, about 5 percent for the disrupted tile system on winter spray, and only 0.4 percent of the time on the oldfield watershed (Table IV). The oldfield watershed received lower input concentrations as will be discussed below. Even so, export of various constituents in runoff remain very low compared to inputs (less than 10% in most cases), and do not represent significant losses (Table III).

The oldfield watershed received water from a downstream lake during the growing season rather than directly from the

Table IV

Concentrations (mg/l) of Nitrogen, Phosphorus, and Chloride in Runoff From Three Oldfield Sites From October 1, 1978, to September 30, 1979. The First Value Listed is the Mean of All Values Measured ± One Standard Deviation; the Second Value is an Unbiased, Flow Weighted Mean Concentration (Calculated by the Method of the International Joint Commission, 1977).

	Oldfield Plots	*Oldfield Watershed*	*Winter Spray Oldfield*
Inorganic N	1.606 ± 1.855 (0.78)*	0.177 ± 0.212 (0)*	0.614 ± 1.227 (0.41)*
	3.231	0.154	1.08
Ammonia N	0.282 ± 1.312	0.065 ± 0.076	0.197 ± 1.089
	0.230	0.069	0.188
Nitrate N	1.232 ± 1.275 (0)*	0.089 ± 0.199 (0)*	0.377 ± 0.536 (0)*
	2.832	0.062	0.850
Nitrite N	0.086 ± 0.141	0.023 ± 0.041	0.040 ± 0.058
	0.167	0.022	0.059
Organic N	0.976 ± 0.798	0.969 ± 0.467	0.918 ± 0.316
	1.367	1.039	0.902
Total P	0.445 ± 0.518 (10.57)*	0.195 ± 0.185 (0.39)	0.367 ± 0.312 (4.63)*
	1.044	0.185	0.486
Soluble Reactive P	0.306 ± 0.421 (6.90)*	0.099 ± 0.150 (0)*	0.248 ± 0.229 (1.07)
	0.822	0.086	0.382
Chloride	91.443±28.645	70.371±23.715	104.195±16.137
	93.286	72.405	97.858

*Values in parentheses indicate percent of samples exceeding 10 mg N/1 or 1 mg P/1.

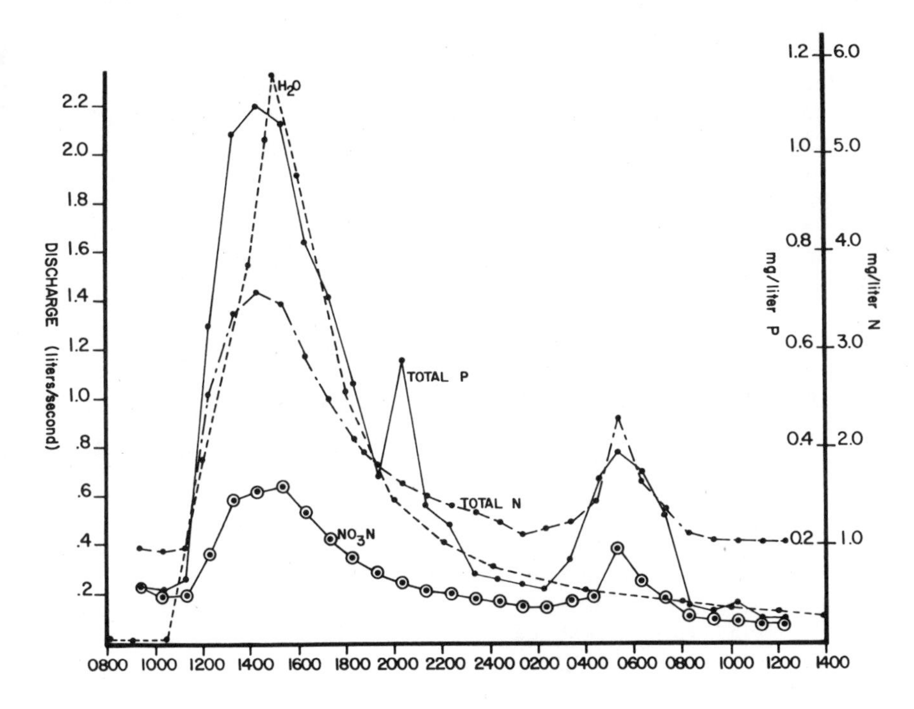

Figure 2. Concentration versus discharge for total P, total N, and NO_3-N for a wastewater irrigation generated runoff event for August 16, 1976 (Hook and Burton, 1978).

East Lansing Sewage Treatment Plant. Thus, input concentrations of inorganic nitrogen were much lower than for the other sites because of the ability of the four lakes to strip nitrogen from wastewater (King and Burton, 1979) and varied from a monthly low of 1.02 ± 0.18 mg N/1 for May to a high of 3.22 ± 0.95 mg N/1 in August with an annual average of 2.09 mg N/1. This low concentration resulted in an application of only 46 kg/ha/yr for this site. Losses from the site are correspondingly low (Table III) even though 55 percent of applied wastewater was lost from this watershed by runoff because of the clay underlying it. Excellent phosphorus renovation also occurred despite this high runoff (Tables III and IV). Water moves through the surface soil to the clay and then laterally to the tile drain before it leaves the site resulting in excellent soil contact time and phosphorus renovation.

Vegetation Responses

Vegetation changes were monitored extensively for the first three years of the study (1975, 1976, 1977). During 1975, changes were monitored for the entire field prior to establishment of the individual plots (F. Reed, unpublished data available from the author). Changes during these first three years are discussed below (from Burton, 1978).

The unharvested vegetation responded to wastewater irrigation by producing more biomass in each of the three seasons of irrigation while the non-irrigated areas produced less biomass in 1976 and 1977 than in the wetter year of 1975 (Figure 3). The effects of wastewater irrigation appeared to be additive with more biomass produced in each succeeding year for the 5 cm/wk area. There was a similar trend for the 10 cm/wk area but the maximum biomass estimates for the two years were not statistically different. Further, there was also a trend for this added biomass to be produced earlier in the growing season with biomass peaking and declining later in the year. This phenomenon was especially pronounced in the third (1977) year of wastewater irrigation (Figure 3). This phenomenon may be explained by two factors. First, added nutrients, especially nitrogen, may be building up in organic matter in the soils over time. This buildup plus the added nutrients in the applied wastewater and the added water itself resulted in increased biomass production. The rate of irrigation had no clear effect on total biomass with the 5 cm/wk treatment resulting in about the same increase as the 10 cm/wk treatment in 1977; but this increase was not as great in the 5 cm/wk area in 1976.

The second factor involves the interaction between individual plant species. In the unharvested plots, goldenrod *(Solidago canadensis* and *Solidago graminifolia)* became more and more dominant during each year of irrigation and accounted for almost all the biomass (Figure 4). Under unirrigated conditions, grasses *(Agropyron repens* and to a lesser extent *Poa compressa)* make up most of the biomass early in the season (until early June). At this time, both *Agropyron repens* and *Taraxacum officinale* (quackgrass and dandelion) flower and their biomass peaks and begins to decline (Figure 5 and 6). This decline coincides with the time that goldenrod *(Solidago)* is entering its most rapid growth phase (Figure 4). Under irrigated, high nutrient conditions the onset of this rapid growth phase appears to be accelerated (Figure 4) and grasses appear to be losing out competitively to goldenrod under these irrigated, non-harvested conditions. This competitive disadvantage appears to be even greater for *Poa compressa* (Canada bluegrass) since it normally peaks later in the season (Figure 7). Thus, irrigation without harvest resulted in greater and greater dominance by goldenrod each

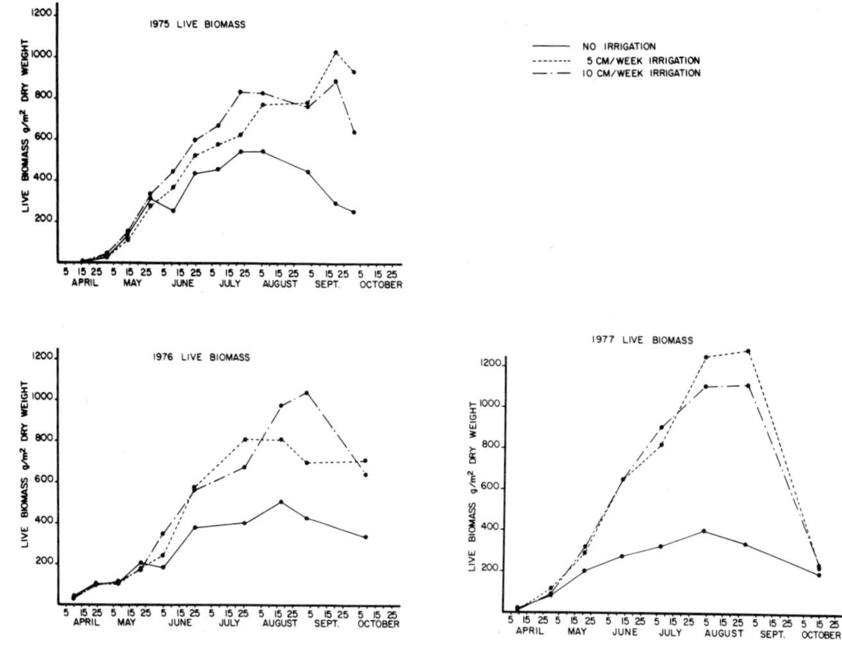

Figure 3. Live biomass accumulation on the unharvested old-field plots. The 1975 data are from unpublished data of F. Reed (Burton, 1978).

year with the grasses becoming less and less dominant. *Aster*, the other major weed species, also was favored under irrigated, unharvested conditions and appeared to be increasing its dominance with time (Figure 8). This increase in dominance appeared late in the season after biomass of goldenrod *(Solidago)* had peaked and started to decline (Figure 4). Thus, this unharvested, wastewater irrigated oldfield is becoming increasingly dominanted by the three weedy species *(Solidago canadensis, Solidago graminifolia,* and *Aster* sp.*)*. Again, the rate of irrigation seems to have little effect on this interaction with the 5 cm/wk and 10 cm/wk areas responding in similar fashion.

Litter biomass decreased on the irrigated plots the first two years compared to the non-irrigated plots but the increased production of 1976 resulted in higher initial litter biomass in 1977 (Figure 9). This litter biomass decreased to levels lower than the non-irrigated plot in July and August but the rapidly declining live biomass showed up as litter on the last sample date of 1977 resulting in much higher litter values in the irrigated plots on that date (Figure 9). Thus, irrigation increased the rate of decomposition but this increase was offset by an increased litter production rate resulting in litter biomass being similar in

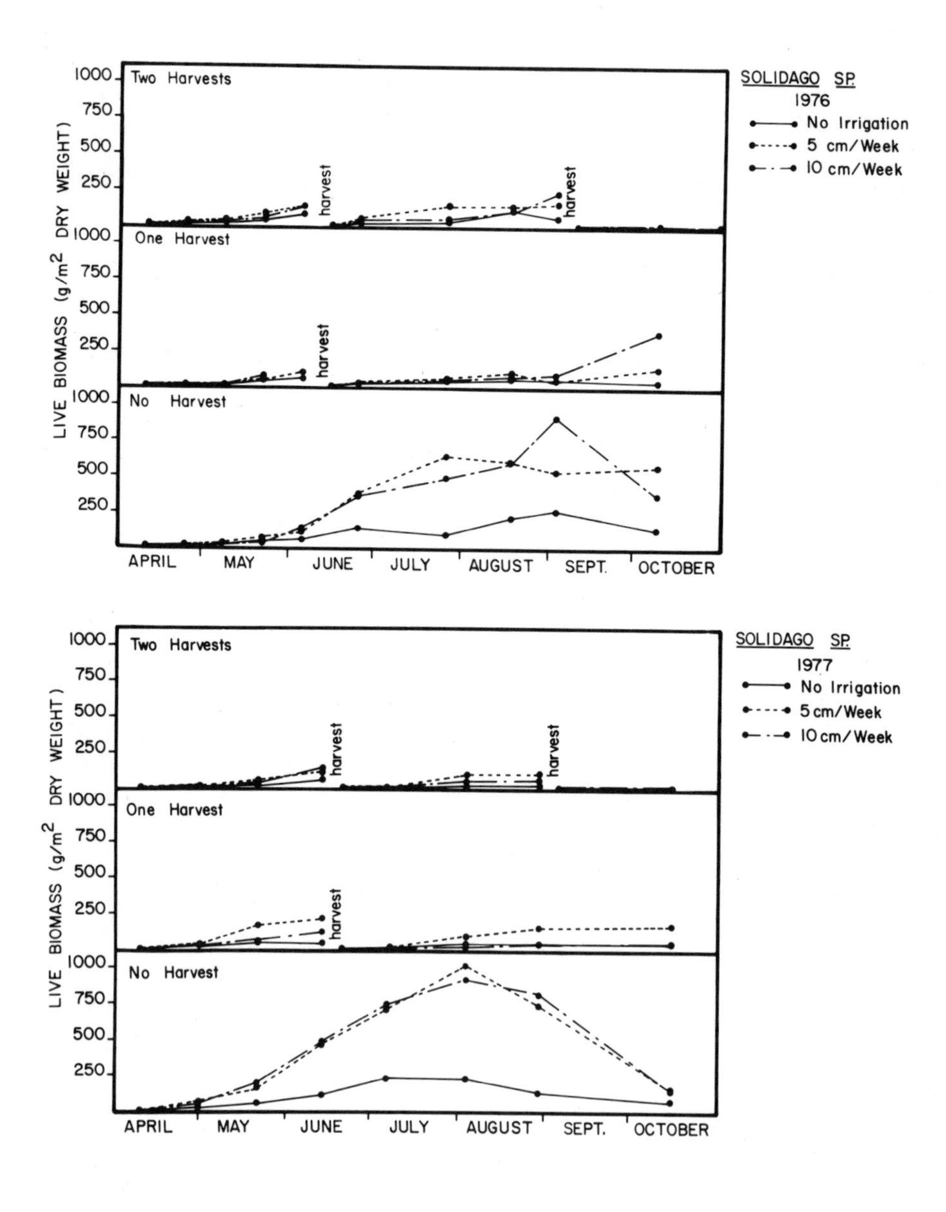

Figure 4. Biomass accumulation by *Solidago* *sp. in response to the various harvest and irrigation treatments (Burton, 1978).*

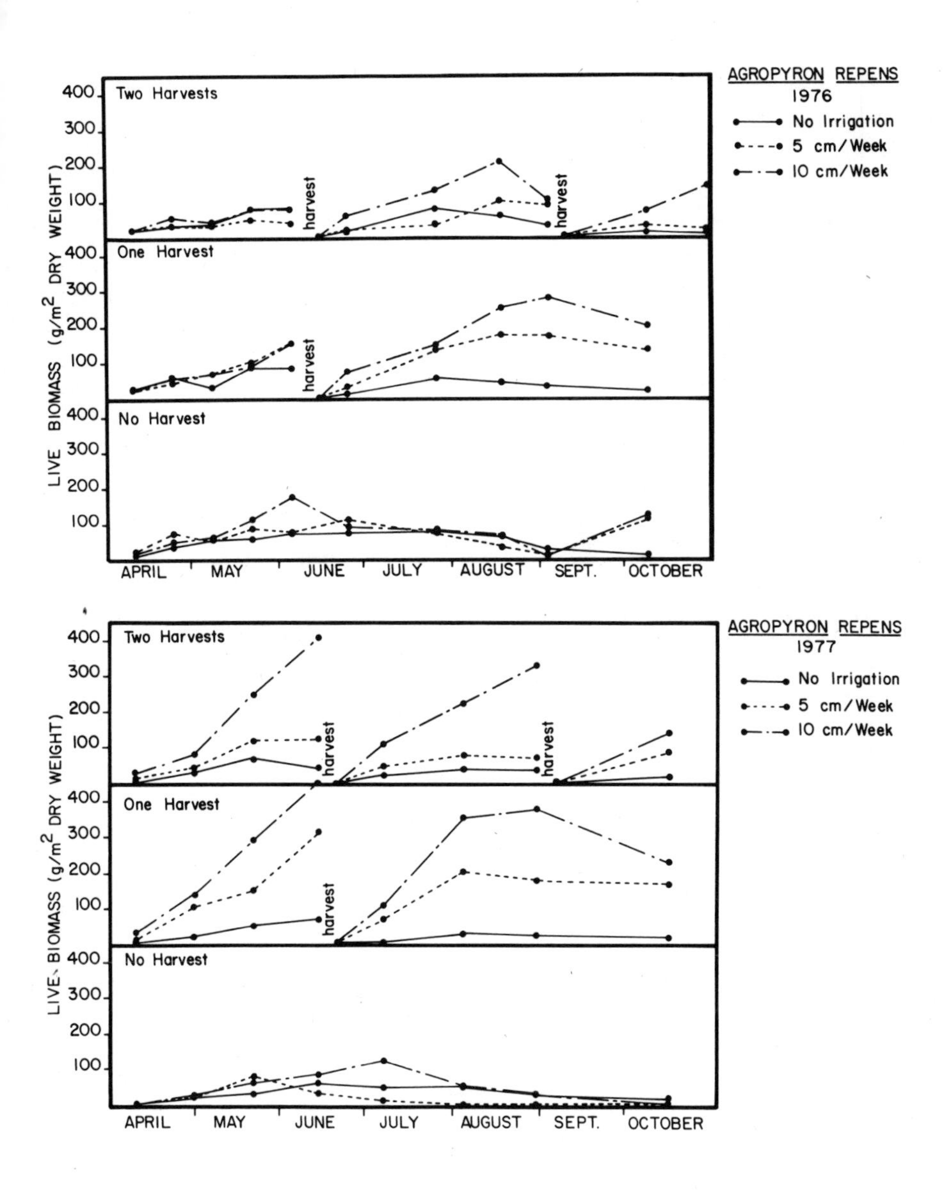

Figure 5. Biomass accumulation by Agropyron repens in response to the various harvest and irrigation treatments (Burton, 1978).

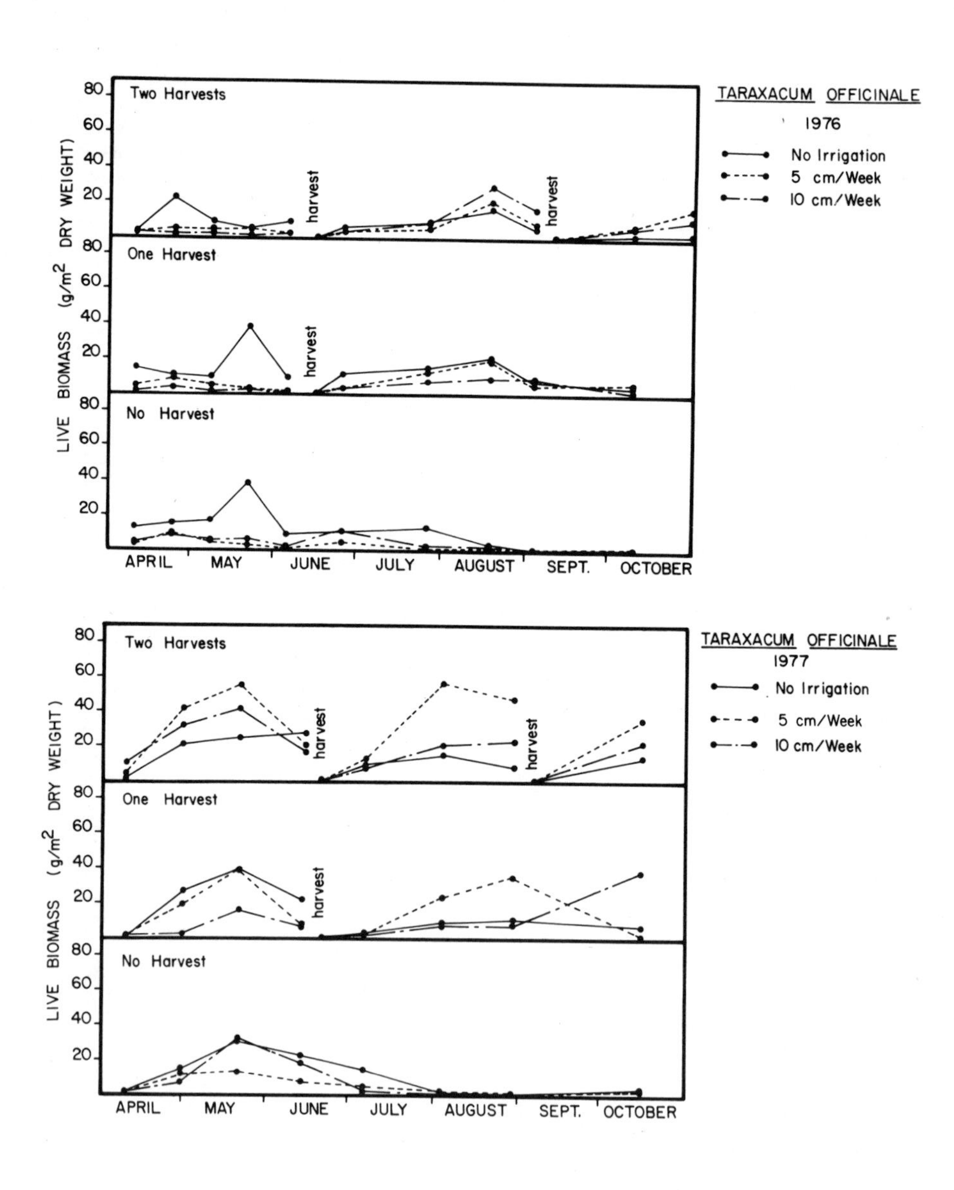

Figure 6. Biomass accumulation by Taraxacum officinale in response to the various harvest and irrigation treatments (Burton, 1978).

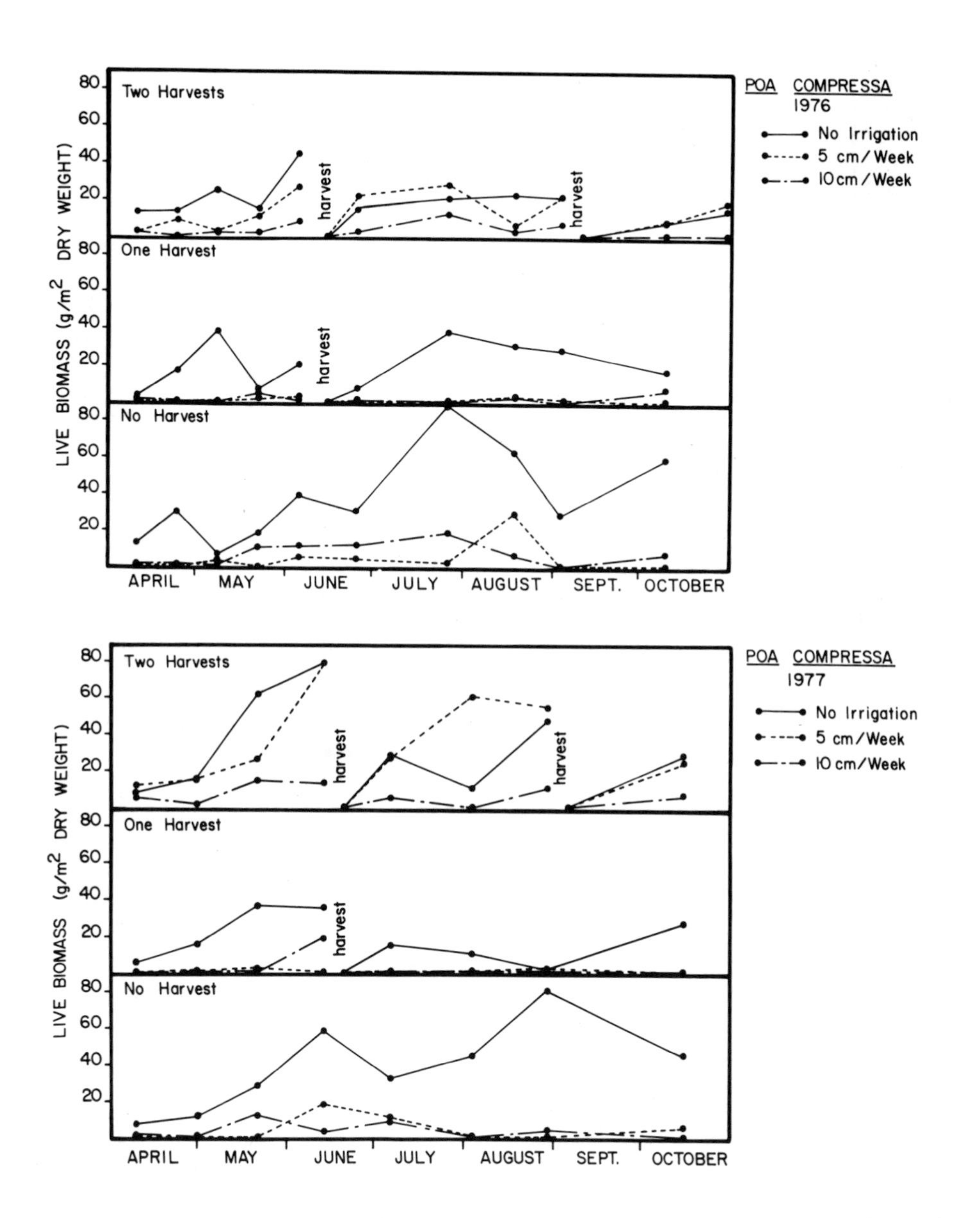

Figure 7. Biomass accumulation by Poa compressa in response to the various harvest and irrigation treatments (Burton, 1978).

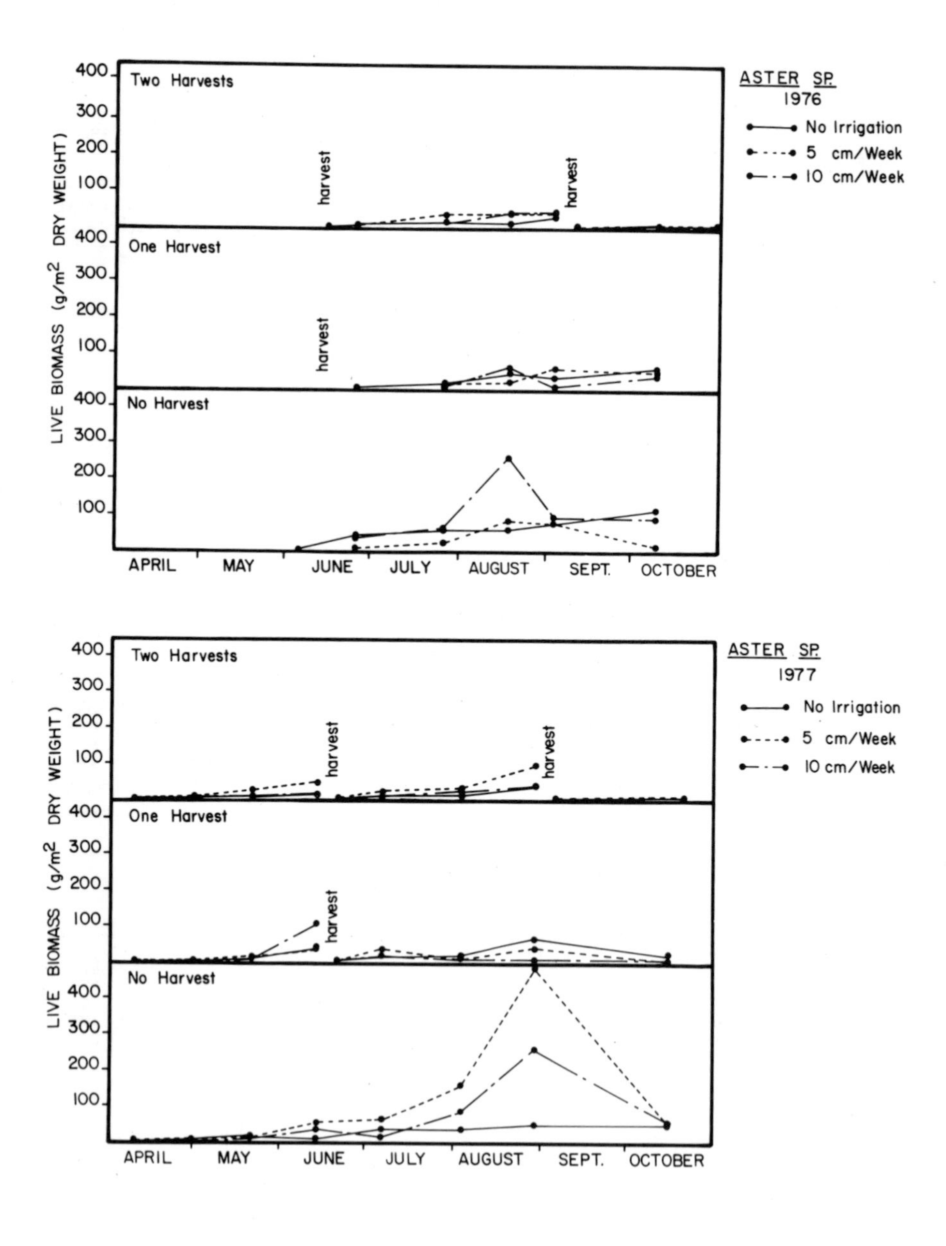

Figure 8. Biomass accumulation of _Aster_ *sp. in response to the various harvest and irrigation treatments (Burton, 1978).*

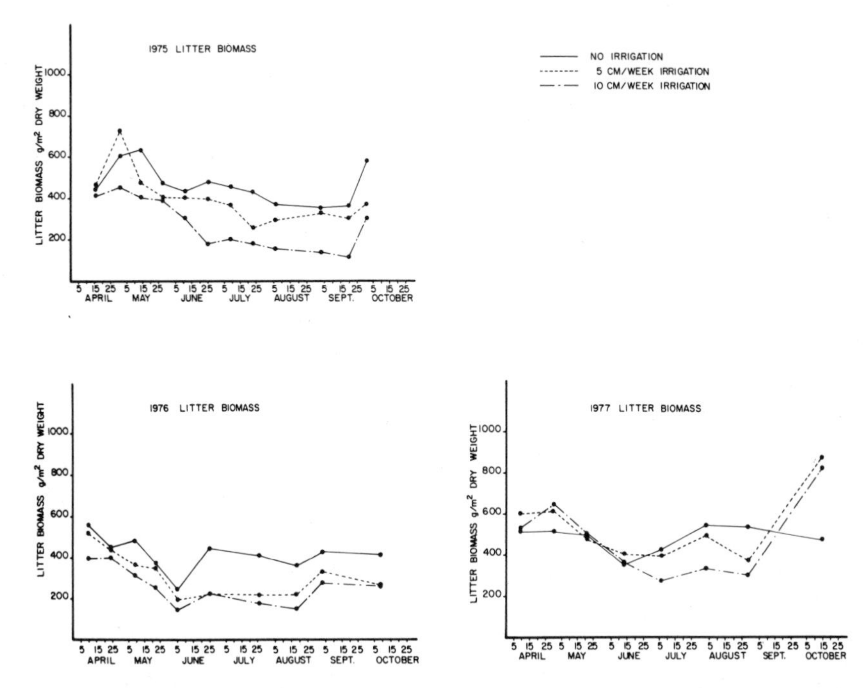

Figure 9. Litter biomass on the unharvested oldfield plots. The 1975 data are from unpublished data of F. Reed (Burton, 1978).

both the irrigated and unirrigated areas on many of the sample dates. This increased litter production and turnover rate is an important mechanism for immobilizing easily leaching nutrients such as nitrogen and releasing them at rates slow enough that they can be recycled into new organic matter production. The increased plant production in successive years of wastewater irrigation, especially on the 5 cm/wk wastewater irrigation plots, could be a result of this mechanism. The rate of irrigation seemed to influence the rate of decomposition with the initial early spring litter biomass decreasing at a more rapid rate on the 10 cm/wk plots than on the 5 cm/wk plots (Figure 9). This phenomenon was especially true during the first year of irrigation but less obvious in successive years (Figure 9).

The irrigated plots responded rapidly after the June harvest, and regrowth resulted in new biomass peaks (Figures 10 and 11). Regrowth slowed and biomass accumulation almost ceased on the harvested plots by the time of the second harvest in early September (Figures 10 and 11); this was especially true in 1977 for all treatments. In 1976, biomass accumulation continued through the end of the sampling period on the 10 cm/wk areas on the one harvest plots but peaked by mid-August on the 0 and 5 cm/wk areas (Figure 10). After the

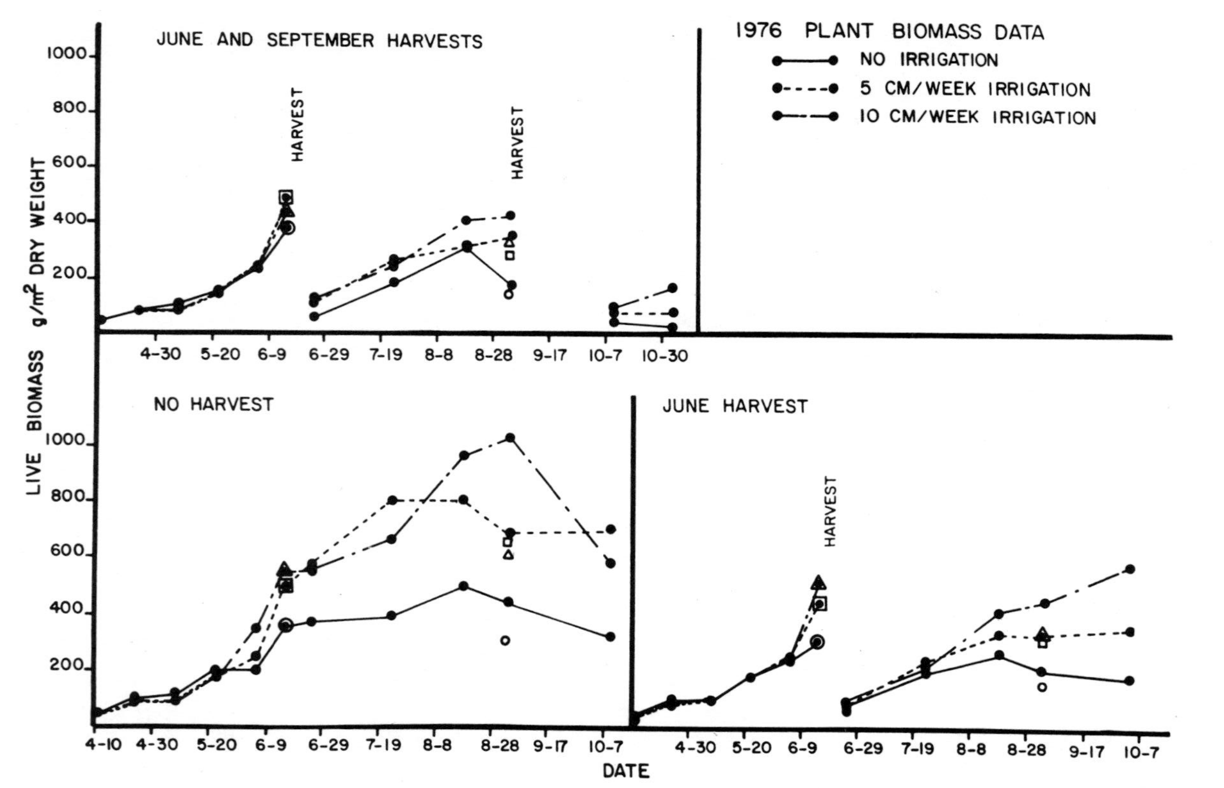

Figure 10. *Total live biomass accumulation in 1976 in response to the various harvest and irrigation treatments (Burton, 1978).*

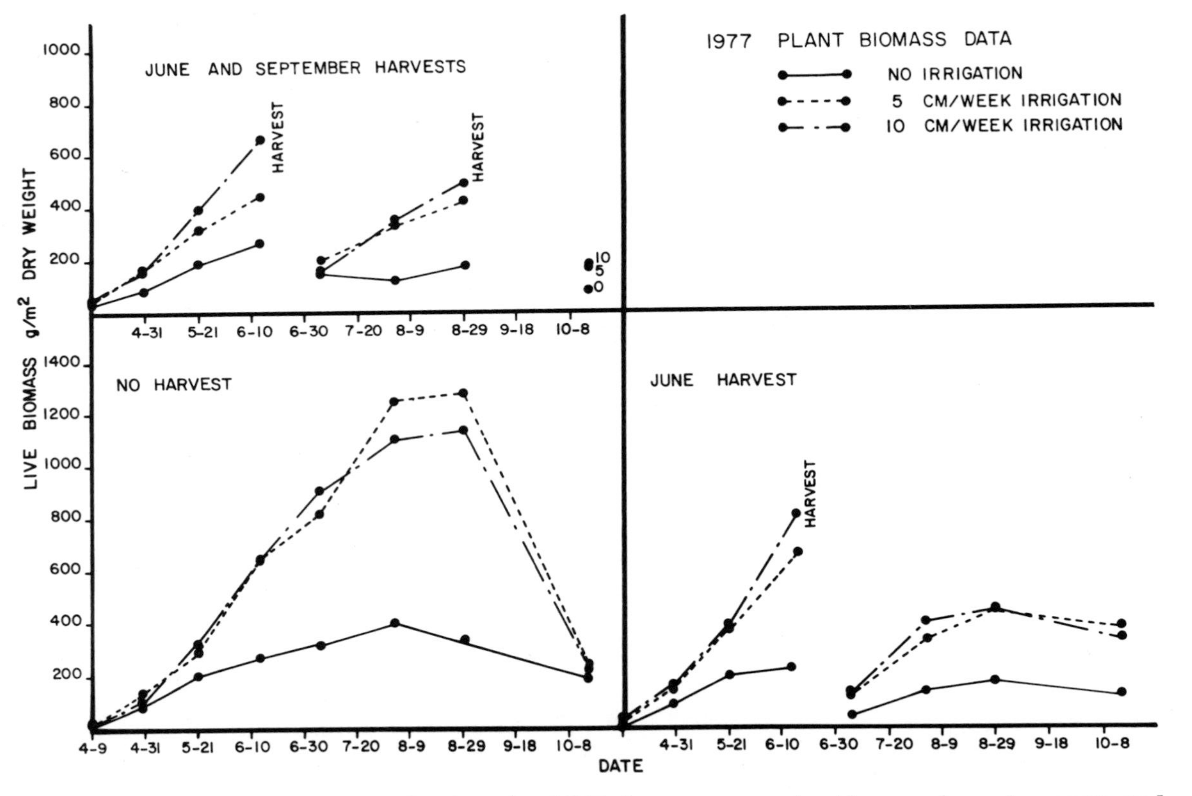

Figure 11. Total live biomass accumulation in 1977 in response to the various harvest and irrigation treatments (Burton, 1978).

second harvest in early September, there was some regrowth with peak biomass estimates of 54 ± 20, 89 ± 35, and 177 ± 97 g/m^2 dry weight for the 0, 5, and 10 cm/wk areas in 1977. Thus, the second harvest promoted growth through October while biomass accumulation essentially ceased on the no harvest and one harvest plots by mid-August to early September. This active growth and nutrient uptake was very important to wastewater renovation since plant uptake of nutrients was extended for an additional 6 weeks each year.

The total or maximum biomass produced appeared to be the same within each irrigation treatment in a given year regardless of harvest treatment (Table V). Harvesting did not appear to consistently increase or decrease the total biomass production for a given irrigation rate. Harvesting did provide three benefits for wastewater treatment. First, it removed accumulated plant nutrients for reuse elsewhere. Second, it prolonged the period of active plant uptake each season. Third, harvesting promoted changes in dominance from the weedy species to the grasses which are more readily utilized as animal feeds. This third phenomenon will be examined in more detail below.

As has already been discussed, irrigation without harvest resulted in greater and greater dominance by the weedy species *(Solidago canadensis, Solidago graminifolia,* and *Aster* sp.*)*, perhaps by accelerating the onset and rate of growth to the point that the grasses *(Agropyron repens* and *Poa compressa)* were shaded out early in the season. Harvesting promoted just the opposite effect. While *Solidago* and *Aster* did regrow after harvest (Figures 4 and 8), they never achieved the large biomass accumulation that they did before harvest. Quackgrass *(Agropyron repens)* regrew rapidly and achieved greater and greater dominance on the harvested plots (Figure 5). This trend was not obvious for *Poa compressa* (Figure 7). *Poa* seems to do best under unharvested, unirrigated conditions. It lost out to *Solidago* and *Aster* under unharvested, irrigated conditions and to *Agropyron* under harvested, irrigated conditions. It did seem to be favored somewhat by a second harvest under all irrigation rates. Thus, *Poa compressa* remained an important minor species on all the unirrigated plots and on the two harvests, 5 cm/wk plots, but tended to lose out on both the once harvested and unharvested plots under irrigated conditions.

The dandelion, *Taraxacum officinale*, under unharvested conditions was an important minor species that was responsible for a significant percentage of plant biomass in April and early May (Figure 6). It grew rapidly early, peaked in early May, and declined to very low biomass levels later in the season. Under harvested conditions, *Taraxacum* regrew rapidly and became an important component of biomass, especially during the second year of harvest on the two harvest

Table V

Maximum Plant Biomass Produced (g/m^2 Dry Weight) on the Oldfield Plots. The Data Represents the Amount Harvested Plus Maximum Biomass Produced After Harvest for Harvested Plots (Burton, 1978).

Irrigation Rate	No Harvest 1976	No Harvest 1977	One Harvest 1976	One Harvest 1977	Two Harvests 1976	Two Harvests 1977
0 cm/wk	503±124	407± 81	587± 78	410± 45	578± 59	533± 58
5 cm/wk	806±210	1291±232	812± 41	1121± 74	869± 82	1055± 93
10 cm/wk	1034±210	1122±493	1107±208	1260±152	965±104	1356±190

treatment (Figure 6). Like *Poa compressa*, dandelion appeared to do best on the harvested plots on the two harvest, 5 cm/wk areas (Figure 6). Thus, both species are influenced by their interactions with the *Solidago* and *Aster* on the unharvested plots and by their interaction with *Agropyron* on the harvested plot. Quackgrass, *Agropyron repens*, produced its greatest biomass on the 10 cm/wk, harvested areas (Figure 5) and appears to be shading out the two minor species as the biomass levels accumulated under such conditions. A second harvest allowed the two minor species to compete more successfully. Regrowth of *Agropyron*, while substantial, was less on the 5 cm/wk areas than on the 10 cm/wk areas (Figure 5) and *Taraxacum* and *Poa* achieved their best growth rate on the 5 cm/wk, two harvest areas (Figures 6 and 7).

There were some 20 to 30 other species that occurred in this oldfield, but these species were fairly rare and unimportant in total biomass production. Some species with clumped distribution were important on an individual sample basis. These species included *Phalaris arundinaceae*, reed canary grass, which grew in clumps in wetter areas; *Cirsium arvense*, Canada thistle; *Phleum pratense*, timothy; and *Rubus* sp. Other fairly common species included *Rumex crispus* and *Rumex acetocella*, *Linaria vulgaris*, *Barbarea vulgaris*, *Daucus carota*, *Plantago* sp., *Potentilla recta*, and *Lactuca canadensis*. *Setaria* sp. was able to invade and become fairly common on the harvested plots, whereas it had not occurred on these plots prior to harvest.

Plant interactions are influenced by both harvest and irrigation. These interactions can be summarized as follows. On untreated control plots, several seasonal biomass peaks account for total plant production. In the early spring, the grasses *(Agropyron repens* and *Poa compressa)* plus the dandelion *(Taraxacum officinale)* dominate growth. *Agropyron repens* is the dominant plant species during the April to early June period. Goldenrod *(Solidago canadensis* and *Solidago graminifolia)* grow slowly during this early period but begin to dominate biomass production by early June. *Solidago* reaches its maximum biomass in July and August with *Solidago canadensis* being the dominant species. *Aster* sp. overlaps *Solidago* with some tendency to reach its biomass peak shortly after *Solidago* in mid-August to September. Under irrigated, non-harvested conditions biomass increases two- to three-fold with almost all of this biomass increase accounted for by *Solidago* and *Aster*. *Solidago canadensis* is the dominant species. Under harvested conditions, total biomass production is about the same as under unharvested conditions but almost all the regrowth after harvest is produced by quackgrass *(Agropyron repens)* with lesser increases by dandelion *(Taraxacum officinale)* and Canada bluegrass *(Poa compressa)*.

The plots were mowed but not harvested in 1978, 1979, and 1980. No intensive studies of community structure have been conducted since mowing was initiated so the following are qualitative observations only. Small areas of mowed plots or, in some cases, most of a mowed plot are presently dominated by Canada thistle *(Cirsium arvense)*. Likewise, wetter plots are dominated by reed canarygrass. However, quackgrass *(Agropyron repens)* continues to dominate most of the mowed area. Likewise, goldenrod *(Solidago* sp.*)* continues to dominate the unharvested plots with aster apparently becoming more common. Thus, trends established during harvest appear to have continued during mowing.

CONCLUSIONS

Oldfield vegetation is very effective at renovation of wastewater. Irrigation with wastewater causes profound changes in community structure. In effect, biomass production increases as nitrogen builds up on the site to biomass peaks 2 to 3 times greater than those of unirrigated sites (from about 400 to 1200 g dry wt/m^2/yr). This increased biomass production is accompanied by lower diversity with only a few species becoming very dominant. The dominant species on unharvested plots are goldenrod *(Solidago* sp.*)* and aster *(Aster* sp.*)*, while the dominant species on harvested plot is quackgrass *(Agropyron repens)*.

Oldfields are effective at nitrogen and phosphorus removal from wastewater and offer excellent tertiary treatment at application rates of up to 10 cm/wk. This nitrogen and phosphorus removal is accomplished through most of the growing season (April-August) with or without mowing or harvest for the first 6 years of operation at least. Harvesting vegetation twice removes the greatest amount of nitrogen and phosphorus in biomass and results in active uptake well into October. Mowing also prolongs the period of active plant uptake of nitrogen and phosphorus.

Nitrogen and phosphorus leachate as a percentage of wastewater input has trended upward over time since mowing only was initiated. Even though wastewater standards are still being met for leachate water quality, this trend does suggest breakthrough at a future date if harvest is not initiated again.

RECOMMENDATIONS

Oldfields offer excellent sites for renovation of wastewater for short periods of 5-6 years with little need for any maintenance other than mowing. Thus, land application sites could utilize such systems for treatment of wastewater while slowly converting these systems to managed fields of perennial forage grasses such as those discussed by M. B. Tesar and others at this conference. In the long-term, oldfields may start to lose nitrate at levels exceeding the 10 mg N/l standard unless they are harvested. This trend appears to be occurring on the WQMF after 6 years but breakthrough, if it occurs, is still to come. Thus, long-term use (more than 5-6 years) of such sites for wastewater renovation would require harvest. Harvested vegetation could be used as a "green manure" or in biomass production of energy. However, the conversion to managed, forage grasses which can be harvested and fed to livestock seems to offer a better long-term solution economically.

ACKNOWLEDGMENTS

This research was supported by Grant A-091-MICH from the U.S. Department of the Interior, Office of Water Research and Technology; by Grant R005143-01 from the U.S. Environmental Protection Agency; and by Michigan State University.

We are indebted to Charles Annett, Paul Bent, Joe Ervin, Scott Farley, John Fogl, Don Herrington, Bobby Holder, and Dan O'Neill for field and technical assistance. A very special note of thanks is due William Baker for his supervision of plant sampling, sorting, and analyses. In addition, many part-time student assistants assisted with plant sorting and are due special thanks. We also thank John Przybyla and Dale Brandes for computer assistance.

LITERATURE CITED

Burton, T.M. 1978. A Mass Balance Study of Recycling Secondary Municipal Wastewater on Abandoned Field Ecosystems. Final Completion Report, Project A-091-MICH, U.S. Department of Interior, Office of Water Research and Technology, Washington, DC 20240 (available N.T.I.S., Springfield, VA 22161, as PB-286275/AS).

Burton, T.M. 1979. Land Application Studies on the Water Quality Management Facility at Michigan State University. In Proc. Second Annual Conference on Applied Research and Practice on Municipal and Industrial Waste, September 18-21, Madison, WI, pp. 112-128.

Burton, T.M. and J.E. Hook. 1978a. Use of Natural Terrestrial Vegetation for Renovation of Wastewater in Michigan. In H.L. McKim (ed.), State of Knowledge in Land Treatment of Wastewater, U.S. Army Cold Regions Research Laboratory, Hanover, NH 03755, August 20-25, 2:199-206.

Burton, T.M. and J.E. Hook. 1978b. Baseline Oldfield Watershed Studies. In T.M. Burton, The Felton-Herron Creek, Mill Creek Pilot Watershed Study. EPA-905/9-78-002, U.S. Environmental Protection Agency, Region V, Chicago, IL 60604, pp. 140-153.

Burton, T.M. and J.E. Hook. 1979. Non-point Source Pollution from Abandoned Agricultural Land in the Great Lakes Basin. J. Great Lakes Res., Int'l. Assoc. Great Lakes Res. 5:99-104.

Burton, T.M. and D.L. King. 1979. A Lake-Land System for Recycling Municipal Wastewater. In Proc. 1979 National Conference Environmental Engineering, July 9-11, ASCE, San Francisco, CA, pp. 68-75.

Burton, T.M. and D.L. King. 1980. The Michigan State University Water Quality Management Facility - A Lake-Land System to Recycle Wastewater. Presented at the International Conference on the Cooperative Research Needs for the Renovation and Reuse of Municipal Wastewater in Agriculture, Hotel Hacienda Cocoyoc, Morelos, Mexico, Dec. 15-19, 1980.

Hook, J.E. and T.M. Burton. 1978. Land Application of Municipal Effluent on Oldfields and Grasslands. In T.M. Burton, The Felton-Herron Creek, Mill Creek Pilot Watershed Study. EPA-905/9-78-002, U.S. Environmental Protection Agency, Region V, Chicago, IL 60604, pp. 25-65.

Hook, J.E. and T.M. Burton. 1979. Nitrate Leaching From Sewage-Irrigated Perennials as Affected by Cutting Management. J. Environ. Qual. 8:496-502.

International Joint Commission. 1977. Quality Control Handbook for Pilot Watershed Studies. International Joint Commission, Windsor, Ontario.

King, D.L. and T.M. Burton. 1979. A Combination of Aquatic and Terrestrial Ecosystems for Maximal Reuse of Domestic Wastewater. In Proc. Vol. 1, Water Reuse - From Research to Application, March 25-30, Washington, DC. AWWA Research Foundation, Denver, CO 80235, pp. 714-726.

Leland, D.E., D.C. Wiggert, and T.M. Burton. 1979. Winter Spray Irrigation of Secondary Municipal Effluent. J. Water Poll. Control Fed. 51:1850-1858.

Thornthwaite, D.W. and J.R. Mather. 1967. Instructions and Tables for Computing Potential Evapotranspiration and the Water Balance. Climatology 10:185-311. (Laboratory of Climatology, Drexel Institute of Technology, Centerton, NJ.)

CHAPTER 8

VEGETATION SELECTION AND MANAGEMENT FOR OVERLAND FLOW SYSTEMS

Antonio J. Palazzo, Thomas F. Jenkins, and C. James Martel
U.S. Army Cold Regions Research and Engineering Laboratory
Hanover, New Hampshire 03755

INTRODUCTION

Overland flow is the mode of land application normally considered for wastewater treatment in areas with tight soils. Wastewater is applied to the upper sections of a gentle slope usually planted with grasses, and is allowed to flow as a thin sheet over the soil surface. At the base of the slope the water is collected for discharge. As with slow rate land treatment systems, the wastewater is renovated by a combination of physical, chemical, and biological processes.

Although there is renewed interest in land treatment, very little information is available about plant growth on overland flow systems. Much of our present knowledge on crop growth in these systems is extrapolated from studies of slow rate sites and conventional agriculture.

Although both slow rate and overland flow systems are land treatment processes, they produce significantly different plant growth environments. The three major differences are the direction of water movement, the types of soil within the system, and the residence time of water within the system boundaries. These differences affect both the degree of treatment achieved for various wastewater constituents and the growth of plants. In overland flow the water moves by sheet flow over the soil surface, resulting in little interaction with the soil. Also, the soils are heavier in texture than at slow rate sites, which impedes air movement and plant growth. This usually results in reduced yields. This growth environment is considerably different from most areas where agronomic plants are grown.

Vegetation selection and management for overland flow systems should include plants which establish rapidly, have economic value, and grow well on tight, moist soils. Because these are wet soils that are gently sloping, they are not appropriate for growing row crops or legumes. Therefore, the selection of herbaceous vegetation is limited to forage grasses.

Little information is available on the importance of the vegetation in removing wastewater constituents and providing a financial return from the sale of crops. Grasses which have been reported to have performed well include Italian ryegrass in Australia (McPherson, 1979) and reed canarygrass in Mississippi and Texas (Lee and Peters, 1979; Thornthwaite, 1969). Plant yields have been reported by Lee and Peters (1979) in Mississippi and Palazzo *et al.* (1980) in New Hampshire.

OBJECTIVES

The objectives of this study were to:

1. determine which species of forage grasses should be used on overland flow systems based on their speed of establishment and long-term persistance,
2. obtain data on nutrient uptake by grasses on overland flow systems to be used to develop design criteria for nutrient removal, and
3. determine forage grass yields and quality.

MATERIALS AND METHODS

The overland flow site is located at the U.S. Army Cold Regions Research and Engineering Laboratory (CRREL) in Hanover, New Hampshire, and has been described in detail elsewhere (Jenkins *et al.*, 1978; Martel *et al.*, 1980). Hanover has a mean annual temperature of 7°C and approximately 160 days of below freezing temperatures each year. Annual rainfall and snowfall average 95 and 185 cm/yr, respectively. The site faces south-southwest and the prevailing winds are from the west and northwest.

The overland flow site is 8.8 m by 30.5 m and is graded to a 5 percent slope (Figure 1). It was divided into three equal sections (designated A, B, and C) each measuring 2.9 m by 30.5 m. The soil placed on the slope was classified as Hartland silt loam (Typic Eystic Eutrochrepts). The cation exchange capacity of the soil was 5 meq/100 g; the pH was 7.1. Bulk densities averaged 1.4 g/cc. Underlying the soil at a depth of 15 cm was a 1.0 mm thick rubber membrane, which prevented downward percolation. Crushed stone was placed at the top of the slope to prevent erosion and to allow for an even flow distribution.

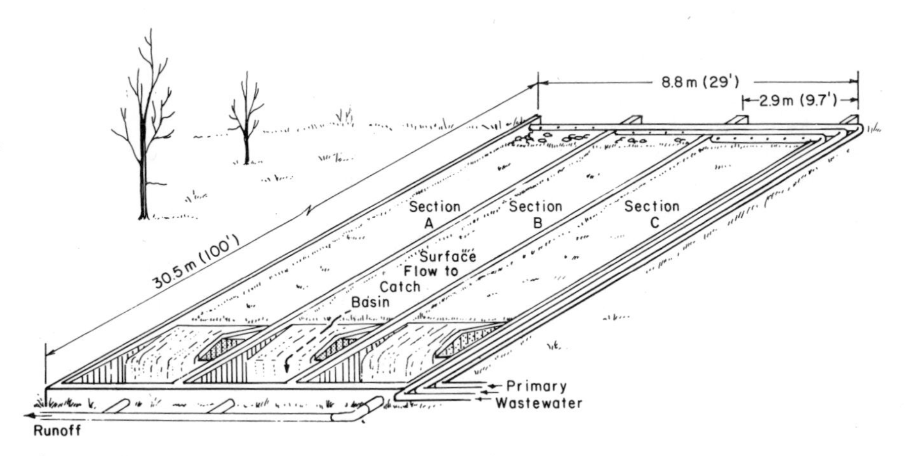

Figure 1. CRREL overland flow research site.

The site was seeded on September 17, 1975, with a seed mixture containing K-31 tall fescue *(Festuca arundinacea* Schreb.*)*, Pennlate orchardgrass *(Phloeum pratense* L.*)*, reed canarygrass *(Phalaris arundinacea* L.*)*, and perennial ryegrass *(Lolium perenne* L.*)* at a rate of 120 kg/ha. The first three grasses were selected for their ability to use nitrogen and to tolerate wet soil conditions. Perennial ryegrass was chosen as a nurse crop to prevent erosion on the slope after seeding.

Prior to seeding in 1975, we applied the equivalent of 2.3 tonne/ha of dolomitic limestone and 726 kg/ha of 15-15-15 grade fertilizer to each slope. The fertilizer supplied 109, 47, and 90 kg/ha of nitrogen, phosphorus, and potassium, respectively. During 1976, the components of the system were tested. Due to the small amounts of nutrients and water applied in 1976, grass growth was poor in the spring of 1977. Therefore, on April 21, 1977, about six weeks prior to wastewater application, the entire site was fertilized with 55 kg/ha of nitrogen from ammonium nitrate, 220 kg/ha of potassium from potassium chloride, and 2.3 tonne/ha of dolomitic limestone. We applied an additional 110 kg K/ha to Section A and 167 kg K/ha to Section B and C on May 6, 1980. The extra potassium was applied to offset deficiencies of this element in the wastewater. Previous studies in slow rate systems have shown potassium to be a limiting factor for plant growth (Palazzo and Jenkins, 1979). We applied additional nitrogen only when wastewater applications were not anticipated; this consisted of 60 kg/ha of nitrogen on June 14, 1978, and 130 kg/ha of nitrogen on May 6, 1980, both on Section A.

Domestic primary sewage from the town of Hanover, New Hampshire, was applied throughout the year. Wastewater was applied at the top of the slope through perforated plastic pipes which could be back-drained when not in use. Runoff was collected at the base of the slope in large galvanized steel tanks. Flowmeters were used to monitor the volumes of water applied and collected.

Each weekly application was applied over a five day period, seven hours per day. The wastewater was analyzed daily, as reported by Jenkins and Palazzo (1981), to determine the amount of nutrients applied. It was generally found to be nearly neutral in pH and to contain an average of 35 mg/l of total nitrogen, mainly in the ammonium form, and 6 mg/l of total phosphorus. Wastewater application began on June 1, 1977, and ended in September, 1980.

The forages were managed annually on a conventional three harvests per year schedule. The forage grass was cut with a sickle bar mower set at a height of 7.5 cm and then hand-raked and weighed to measure the total fresh weight per section. Subsamples from each plot were dried at 70°C for 48 hr to determine total dry weight yields. Grab samples of the dried grasses were analyzed for nitrogen, phosphorus, potassium, and six trace elements according to methods described by Liegel and Schulte (1977).

RESULTS AND DISCUSSION

Loading Rates

Annual loading rates were determined by totaling the amounts of wastewater applied between harvests. This included all the wastewater applied during winter before the first harvest period. Loading rates during winter were usually high. Therefore, plant uptake would have been more efficient, on a percentage basis, if winter applications during plant dormancy were not included in the total loading rate of wastewater nutrients.

The annual nitrogen and phosphorus loading rates in this study ranged up to 2038 and 228 kg/ha, respectively (Table I). The highest loading rates are probably beyond what would be used for nitrogen and phosphorus treatment in overland flow systems.

Wastewater applications were not consistent in each harvest period (Table I). Years in which sections received significant amounts of wastewater in at least two of the three harvest periods were: 1978 on Section A, 1978 and 1979 on Section B, and 1979 and 1980 on Section C.

Botanical Composition

It is important to maintain a vigorous community of desirable plants on overland flow slopes. If a healthy

Table I

Amounts of Nitrogen and Phosphorus Applied Between Harvests.

		Harvest Periods			
Section	*Year*	*First*	*Second*	*Third*	*Total*
			Nitrogen (kg/ha)		
Section A	1977	*	97	98	195
	1978	655	0	279	934
	1979	331	0	0	331
	1980	0	0	0	0
Section B	1977	*	96	110	206
	1978	587	0	280	867
	1979	354	361	585	1300
	1980	1704	209	113	2026
Section C	1977	*	1	1	2
	1978	2	0	236	238
	1979	336	321	603	1260
	1980	1702	221	115	2038
			Phosphorus (kg/ha)		
Section A	1977	*	17	18	35
	1978	106	0	50	156
	1979	56	0	0	56
	1980	0	0	0	0
Section B	1977	*	19	20	39
	1978	95	0	56	151
	1979	60	70	98	228
	1980	280	12	8	226
Section C	1977	*	2	1	3
	1978	5	0	43	48
	1979	57	61	98	216
	1980	203	13	8	224

*System was not yet operating.

community is not maintained, soil erosion will result in the development of small channels, which reduce wastewater treatment efficiency. If desirable species are replaced by weeds, nitrogen and phosphorus uptake will be reduced, resulting in poor wastewater renovation. In addition, if less desirable species predominate, it will reduce the market value of the crop or necessitate costly re-seeding operations and hence increase overall operation and management costs.

Of the four grasses sown, orchardgrass and tall fescue were the most persistant over the initial three years of wastewater application. They were able to germinate well under low nutrient conditions and were able to persist on the slopes. Perennial ryegrass established well initially; however, after two winters of wastewater application, few plants of this species survived. The low survival rate was probably due to winter injury, to which ryegrass is known to be sensitive.

Reed canarygrass did not become well established on the site initially. This grass is slow to germinate and requires large amounts of nutrients and water to grow actively. Since only a small amount of fertilizer was applied during seeding, the soil apparently was not fertile enough for adequate growth once the reed canarygrass was established. After wastewater applications began in 1977, the growth of reed canarygrass improved steadily; it became one of the dominant species on the site during the fourth season. This experiment demonstrated that commercial fertilizers should be applied at low rates (50 kg/ha each of nitrogen, phosphorus, and potassium) to overland flow slopes after the system has been constructed if delays in wastewater application are expected. The fertilizer will favor a healthy grass stand and will result in better performance once the system goes into operation.

Other plants, such as Kentucky bluegrass and quackgrass, invaded the site as the study progressed. These species have been observed to remove large amounts of nutrients in experiments on slow rate systems (Palazzo and McKim, 1978; Marten *et al.*, 1979) and are, therefore, desirable. Kentucky bluegrass is an excellent forage grass and appears to be tolerant to wet soils. Quackgrass, considered a poor forage species for animal feed under conventional agricultural management, has been observed to be of excellent quality when irrigated with wastewater (Marten *et al.*, 1978). Therefore, both of these species can be acceptable components of the grass mixture on overland flow slopes.

Barnyardgrass *(Echinochloa crusgualli* [L.] Beaub*)*, another invading species, became dominant at the top of the slopes during the study (Figure 2). This probably began when wastewater solids smothered more desirable plants at the top of the slope (Figure 3). Barnyardgrass is an annual which

Figure 2. Top photo shows the tall, aggressive barnyardgrass during summer, 1979, on Section B of the overland flow system. Bottom photo shows the same section the following spring.

Figure 3. Wastewater solids near top of the overland flow slope.

Figure 4. Newly seeded grasses in a barren area of slope.

germinates from seed in the spring and becomes tall and aggressive. During the summer of 1979, it dominated the desirable grasses; but in early fall it died, creating barren areas and leaving the slope susceptible to soil erosion until the following spring (Figure 2). The reduction in plant uptake of nitrogen and phosphorus in 1980, discussed later in this report, was due to barnyardgrass. Since it crowds out desirable plants, barnyardgrass should be controlled. Barren areas should be re-seeded with high-yielding perennial grasses (Figure 4).

At the end of the study (October, 1980), the botanical composition of the three sections was assessed. At this time Section A had not been irrigated with wastewater for 15 months (a low nutrient condition), while Sections B and C had received wastewater on a daily basis. In terms of botanical composition, the site could be divided into three areas: the upper and lower sections of the treated slopes and the untreated low nutrient area.

In the upper portions of Sections B and C, where the deposition of solids was greatest, the grasses consisted primarily of quackgrass, reed canarygrass, and, to a lesser extent, Kentucky bluegrass. The percentage of soil covered by grasses in this area was also not as great as in the lower parts of the slope. The barren areas were largely a result of barnyardgrass dying out. In the lower halves of Sections B and C, the dominant grass was Kentucky bluegrass, with lesser amounts of orchardgrass and trace amounts of reed canarygrass. In Section A, Kentucky bluegrass, orchardgrass, and tall fescue were present. As with perennial ryegrass, tall fescue did not survive after the third season in areas which received wastewater. In all treated sections, the dominant grasses looked healthy and vigorous.

Recent research has shown that BOD and volatile organics may be treated in colder temperatures (Martel *et al.*, 1980; Jenkins *et al.*, 1980). Hence, it may become feasible on some slopes to apply wastewater during winter periods. However, this practice is not recommended if a healthy grass cover is to be maintained, as ice covers or crusts over the grasses which develop due to cold weather applications can lead to winter injury and barren soil areas (Figure 5). This crust could prevent the exchange of gases required for plant metabolism (Dexter, 1956).

The nitrogen-potassium balance in wastewaters may also be insufficient to meet plant needs, leading to poor plant growth and possible winter injury (Palazzo and Jenkins, 1979). The total amount of potassium applied, from either wastewater or fertilizer, should be approximately 90 percent of the amount of nitrogen removed by the crop. Nitrogen and potassium have been called the most prominent combination of elements affecting plant winter hardness, especially for forage grasses (Kresge, 1974).

Figure 5. *Top photo shows ice buildup on the lower part of the slope due to winter applications of wastewater. Bottom photo shows areas barren of vegetation during the following spring.*

Plant Yields and Analysis

Annual plant yields on a dry weight basis, when wastewater was applied for a minimum of two harvest periods, ranged between 7.6 and 12.2 tonne/ha (Table II). The mean yield was 9.7 tonne/ha. Yields ranged from 1.1 to 7.4 tonne/ha when little or no wastewater was applied.

Table II

Plant Yields by Harvest Periods at the Overland Flow Test Site.

	Harvest Period			
Year	*First*	*Second*	*Third*	*Total*
		Section A (tonne)		
1977	-	1.8	2.0	3.7**
1978	3.6	3.6*	1.5	8.7
1979	4.1	1.4*	1.9*	7.4**
1980	2.0*	0.8*	0.4*	3.3**
		Section B (tonne)		
1977	-	1.6	1.9	3.5**
1978	2.9	5.0*	1.5	9.5
1979	4.6	4.1	3.4	12.2
1980	2.0	3.5	2.1	7.6
		Section C (tonne)		
1977	-	0.5*	0.6*	1.1**
1978	0.3*	2.7*	2.0	4.9**
1979	3.2	3.7	3.8	10.6
1980	3.7	2.3	3.5	9.5
Mean	2.9	2.6	2.1	

*Harvest periods when wastewater was not applied.
**Growing seasons in which wastewater was not applied during two or more harvest periods.

The average yield as a result of wastewater application was almost three times the average hay yield in New Hampshire (NECLRB, 1978) and was only slightly lower than that produced at the adjacent CRREL slow rate test site. At the slow rate site, yields of a mixed forage grass averaged 13.1 tonne/ha and ranged between 7.4 and 17.6 tonne/ha over a seven year period (Jenkins and Palazzo, 1981).

The value of the hay produced in this study was calculated by adding 15 percent to the dry weight yields to account for the normal moisture content of hay. The average hay yield, plus 15 percent for moisture, was 11.2 tonne/ha or 11,200 kg/ha. If hay is worth \$77/tonne (\$70/ton), the annual value of one hectare of hay is \$862 (\$349/acre).

Plant yields increased at higher annual application rates of nitrogen up to about 1200 kg N/ha (Figure 6). At applications of about 2000 kg/ha of nitrogen, yields declined. The reason for the decline at this high application rate is unclear. On Section B, which yielded about 7600 kg/ha in 1980, the decline appears to be related to the dominance of barnyardgrass on the site the year before. This plant left large barren areas, which reduced overall plant yields. The decline could also be related to the large amounts of water applied during this period; this could have injured plants or reduced the depth of aerated soil.

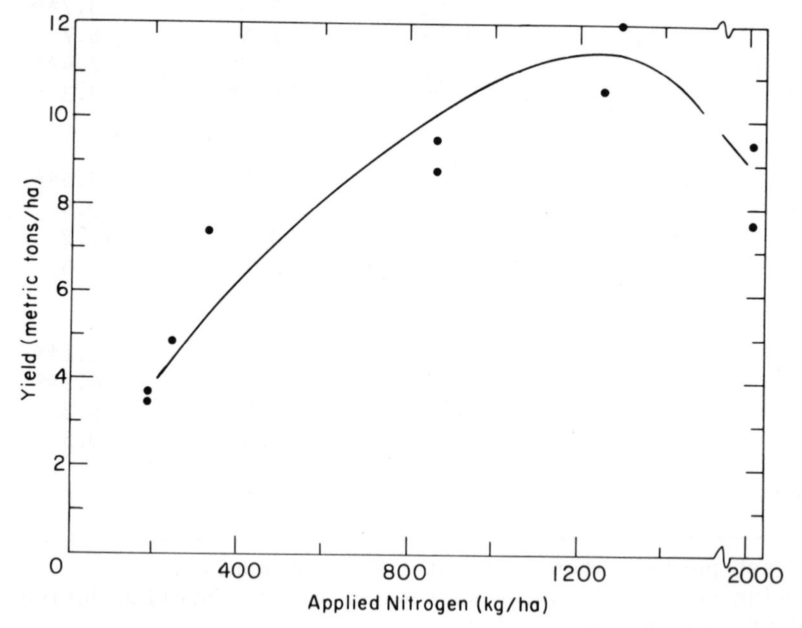

Figure 6. Annual plant yields under continuous nitrogen loading rates throughout the year.

Yields were highest in the first harvest period; they declined slightly in the second and third periods (Table II). The mean daily growth of plants receiving wastewater was far greater during the first harvest period: 96, 50, and

34 kg/ha/day for the first, second, and third harvest periods, respectively.

A similar trend for mean daily yields was reported for a two-year study at an adjacent slow rate site (Palazzo and Graham, 1981). The actual mean daily yields on the slow rate system, however, were found to be slightly greater than at the overland flow site; they were 106, 75, and 59 kg/ha/day during the first, second, and third periods, respectively.

The forage grasses from Sections B and C were analyzed for feed quality during the last cuttings in 1979 and 1980 (Table III). They contained an average of 23 percent crude protein and 85 percent total digestible nutrients. Barnes (1975), discussing the standards for grass hay, noted that grass testing 15 percent or above in crude protein and 65 percent or above for total digestible nutrients is considered excellent in quality.

Table III

Feed Quality of Forage Grasses Receiving Wastewater and Reported Standards for High Quality Feed.

Harvest Date	*Section*	*Crude Protein (Percent)*	*TDN* (Percent)*
Forage Grasses			
October 15, 1979	B	18	69
September 30, 1980	B	26	101
October 15, 1979	C	21	69
September 30, 1980	C	24	99
Mean	-	23	85
Reported Standards			
Barr and Staubus (1980)		15-16	70-75
Barnes (1975)		>18	>65

*TDN = Total Digestible Nutrients

The plants were analyzed for nine elements at each harvest (Table IV). None of the concentrations of the elements measured was in the deficiency or toxicity ranges, indicating that the plants were healthy under the management schemes used in this study.

Table IV

Mean Plant Composition.

Element	*Concentration*
Nitrogen	2.83%
Phosphorus	0.41%
Potassium	2.45%
Calcium	0.53%
Magnesium	0.25%
Manganese	81 mg/kg
Zinc	36.0 mg/kg
Copper	9.1 mg/kg
Boron	5.4 mg/kg

Plant Uptake of Nitrogen and Phosphorus

When wastewater was applied for a minimum of two harvest periods, annual plant uptake ranged from 210 to 332 kg N/ha and from 27 to 48 kg P/ha (Table V). During seasons with limited applications of wastewater, plant uptake was lower; for nitrogen and phosphorus ranging between 28 and 168, and 4 and 27 kg/ha, respectively.

Uptake of nutrients increased with increasing loading rates but either leveled off or decreased at the highest loading rate (Figures 7 and 8). The highest loading rates occurred on Sections B and C in 1980. The lower uptake on Section B, which amounted to 106 kg/ha less nitrogen than Section C (Table V), was primarily related to the lack of vegetation on this slope. This section, as previously discussed, had been dominated by barnyardgrass the previous season at the expense of perennial grasses.

Mean daily plant uptake of nitrogen and phosphorus during each harvest period is shown in Table VI. The most rapid uptake of nutrients occurred during the first harvest period when plants incorporated an average of 2.8 and 0.4 kg/ha/day of nitrogen and phosphorus, respectively. The rate of daily uptake was similar to that found in an adjacent slow rate system (Palazzo and Graham, 1981).

CONCLUSIONS AND RECOMMENDATIONS

Plant growth was excellent when wastewater was applied. Reed canarygrass, orchardgrass, and tall fescue had good persistence on the slopes during the study. Perennial ryegrass was dominant early in the study and was effective in removing

Table V

Plant Uptake of Nitrogen and Phosphorus at the Overland Flow Test Site.

Section	Year	Harvest Periods First	Second	Third	Total
			Nitrogen (kg/ha)		
Section A	1977	-	46	61	107**
	1978	118	82*	43	243
	1979	103	36*	29*	168**
	1980	71*	19*	12*	102**
Section B	1977	-	50	41	91**
	1978	84	137*	51	272
	1979	139	108	85	332
	1980	45	96	69	210
Section C	1977	-	14*	14*	28**
	1978	8*	83*	60	151**
	1979	93	108	124	325
	1980	112	69	135	316
			Phosphorus (kg/ha)		
Section A	1977	-	7	8	15**
	1978	19	16*	5	40
	1979	15	5*	7*	27**
	1980	6*	3*	2*	11**
Section B	1977	-	7	8	15**
	1978	15	26*	7	48
	1979	18	17	13	48
	1980	7	12	8	27
Section C	1977	-	2*	2*	4**
	1978	1*	12*	9	22**
	1979	13	18	17	48
	1980	14	9	17	40

*Harvest periods when wastewater was not applied.
**Growing seasons in which wastewater was not applied during two or more harvest periods.

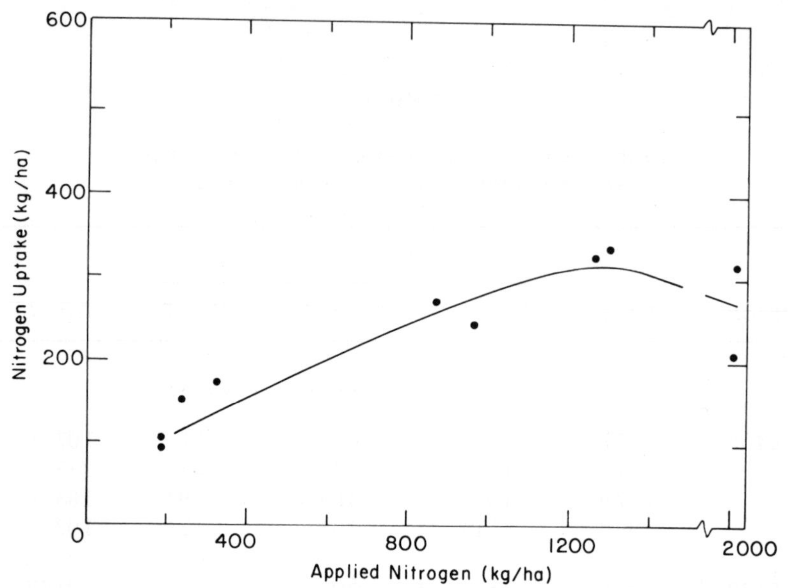

Figure 7. *Annual plant uptake of nitrogen at various nitrogen loading rates.*

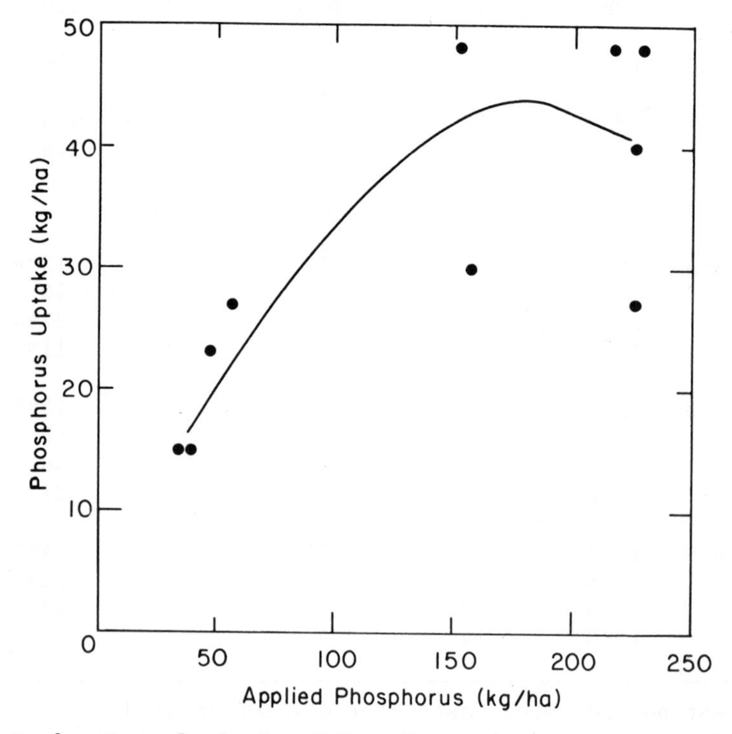

Figure 8. *Annual plant uptake of phosphorus under continuous phosphorus loading rates throughout the year.*

Table VI

Mean Daily Uptake of Nitrogen and Phosphorus on the Overland Flow Slopes.

Harvest Period	*Days of Growth*	*Nitrogen Uptake (kg/ha/day)*	*Phosphorus Uptake (kg/ha/day)*
First (May-June)	36	2.8	0.4
Second (June-July)	57	1.4	0.2
Third (July-September)	71	1.1	0.1

nutrients and preventing soil erosion while the perennial grasses became established. Kentucky bluegrass and quackgrass, both of which invaded the experimental area, were also found to be acceptable species.

Plant selection should be limited to forage grasses. Seed mixtures containing perennial ryegrass plus two or three of the other species should be suitable for overland systems.

Since the terrain is gently sloping, grasses should be established as soon after grading as possible to minimize erosion. An application of fertilizer and irrigation should be considered for more rapid plant establishment.

High plant yields can be expected on overland flow systems. Yields in this study were almost three times higher than the average hay yield in New Hampshire; they ranged from 7.6 to 12.2 tonne/ha when receiving wastewater during the entire growing season. The quality of the forage, in terms of crude protein and total digestible nutrients, was excellent and the mean dry weight yield (9.7 tonne/ha was worth about $862/ha.

Annual plant uptake of nitrogen and phosphorus during seasons of wastewater application ranged between 210 and 332 kg/ha and 27 and 48 kg/ha, respectively. Initially, uptake increased with increasing loading rates but eventually leveled off. The rate of plant uptake was highest during the first harvest period.

Proper management is important in maintaining grass growth and uptake. This study showed that grasses should receive light applications of commercial fertilizer (about

50 kg/ha each of nitrogen, phosphorus, and potassium) if delays in wastewater applications occur after seeding.

Annual weeds should be controlled. In this study, barnyardgrass crowded out the more desirable perennial grasses and reduced plant nutrient uptake on the slope.

Only healthy plants can provide optimum plant growth and nutrient removal. Grasses should, therefore, be periodically checked for their chemical composition to determine nutrient deficiencies or toxicities. Analysis of plants and recommendations based on the results are usually provided at local state agricultural experiment stations. Prior analysis of the wastewater will indicate specific elements of concern. They will include those required by plants for growth or those that may be toxic, such as metals.

Grasses should also be managed to increase tolerance to cold. The amount of potassium applied should be 90 percent of the amount of nitrogen removed by the plant. Again, wastewater should be analyzed to determine the nitrogen-potassium ratio of the wastewater, which will indicate if supplemental potassium applications are required.

Wastewater solids should be controlled so that grass smothering at the point of wastewater application does not occur. Bare areas should be re-seeded with desirable species as soon as possible to prevent weed encroachment and to improve plant uptake.

ACKNOWLEDGMENTS

The authors thank Mr. Sherwood Reed, Dr. C. Edward Clapp, and Mr. David Cate for their review and constructive comments on the manuscript. We also thank Ms. Patricia Schumacher for wastewater analysis information and Mr. John M. Graham and Mr. Carl Diener for technical assistance.

REFERENCES

Barnes, R.F. 1975. Forage Testing and Its Applications. In M.H. Heath (ed.), Forages. Iowa University Press, Ames, IA, pp. 654-663.

Barr, H.L. and J.R. Staubus. 1980. Disappointed in How Your Cows Produce? Hoard's Dairyman 125:1238.

Dexter, S.T. 1956. The Evaluation of Crop Plants for Winter Hardiness. Adv. Agronomy 8:203-239.

Jenkins, T.F., C.J. Martel, D.A. Gaskin, D.J. Fisk, and H.L. McKim. 1978. Performance of Overland Flow Land Treatment in Cold Climates. In Proc. State of Knowledge in Land Treatment of Wastewater, Hanover, NH 03755, August 20-25, 2:61-70.

Jenkins, T.F., D.C. Leggett, and C.J. Martel. 1980. Removal of Volatile Trace Organics From Wastewater by Overland Flow Land Treatment. J. Environ. Sci. Health A15:211-224.

Jenkins, T.F. and A.J. Palazzo. 1981. Wastewater Treatment by a Slow Rate Land Treatment System. CRREL Report. U.S. Army Cold Regions Research and Engineering Laboratory, Hanover, NH 03755.

Kresge, C.B. 1974. Effect of Fertilization on Winter Hardiness of Forages. In D.A. Mays (ed.), Forage Fertilization. American Society of Agronomy, Madison, WI 53706, pp. 437-453.

Lee, C.R. and R.E. Peters. 1979. Overland Flow Treatment of a Municipal Lagoon Effluent for Reduction of Nitrogen, Phosphorus, Heavy Metals, and Coliforms. Progress in Water Technology 11:175-184.

Liegel, E.A. and E.E. Schulte. 1977. Wisconsin Soil Testing, Plant Analysis, and Feed and Forage Analysis Procedures. University of Wisconsin, Madison, WI 53706.

Martel, C.J., T.F. Jenkins, and A.J. Palazzo. 1980. Wastewater Treatment in Cold Regions by Overland Flow. CRREL Report 80-7. U.S. Army Cold Regions Research and Engineering Laboratory, Hanover, NH 03755.

Marten, G.C., C.E. Clapp, and W.E. Larson. 1979. Effects of Municipal Wastewater Effluent and Cutting Management on Persistence and Yield of Eight Perennial Forages. J. Environ. Qual. 8:650-658.

Marten, G.C., R.H. Dowdy, W.E. Larson, and C.E. Clapp. 1978. Feed Quality of Forages Irrigated With Municipal Sewage Effluent. In Proc. State of Knowledge in Land Treatment of Wastewater, Hanover, NH 03755, August 20-25, 2:183-190.

McPherson, J.B. 1979. Land Treatment of Wastewater at Werribee Past, Present, and Future. Progress in Water Technology 11:15-32.

NECLRB. 1978. Milk and Feed. New England Crop and Livestock Reporting Board, Concord, NH, August 16.

Palazzo, A.J. and J.M. Graham. 1981. Seasonal Growth and Uptake of Nutrients by Orchardgrass Irrigated With Wastewater. CRREL Report 81-8. U.S. Army Cold Regions Research and Engineering Laboratory, Hanover, NH 03755.

Palazzo, A.J. and T.F. Jenkins. 1979. Land Application of Wastewater: Effects on Soil and Plant Potassium. J. Environ. Qual. 8:309-312.

Palazzo, A.J., C.J. Martel, and T.F. Jenkins. 1980. Forage Grass Growth on Overland Flow Systems. In Proc. of 1980 ASCE National Conference on Environmental Engineering, July, New York, NY, 8 pp.

Palazzo, A.J. and H.L. McKim. 1978. The Growth and Nutrient Uptake of Forage Grasses When Receiving Various Application Rates of Wastewater. In Proc. State of Knowledge in Land Treatment of Wastewater, Hanover, NH 03755, August 20-25, 2:157-163.

C.W. Thornthwaite Associates. 1969. An Evaluation of Cannery Waste Disposal of Overland Flow Spray Irrigation. Laboratory of Climatology, Publication in Climatology, Elmer, NJ, Volume 22, Number 2, 73 pp.

CHAPTER 9

GROWING TREES ON EFFLUENT IRRIGATION SITES WITH SAND SOILS IN THE UPPER MIDWEST

John H. Cooley
North Central Forest Experiment Station
Forest Service, U.S. Department of Agriculture
1407 S. Harrison Road
East Lansing, Michigan 48824

INTRODUCTION

In the United States, the earliest examples of wastewater application to planted trees and established forests involved cannery waste disposal. The wastewater generally hindered the trees' survival or growth or both because application rates were not matched with the hydraulic characteristics of the soils or the moisture and nutrient requirements of the trees (Little *et al.*, 1959). Nor did growers recognize the unique cultural requirements for establishing trees on irrigated sites (Rudolph, 1957; Rudolph and Dils, 1955).

More recently, much lighter dosages of effluent from mechanical treatment plants and oxidation ponds have been applied to a variety of newly planted and established stands. At two locations in lower Michigan, several kinds of trees with a wide range of nutrient and moisture requirements have been irrigated with oxidation pond effluent during their establishment and initial growth stages. The effects are summarized in Table I. A pole-size unthinned red pine plantation has also been irrigated, in addition to stands of thinned and unthinned pole-size northern hardwoods. Several species not tried in Michigan and some site conditions not represented here have been evaluated elsewhere in the eastern United States (Sopper and Kardos, 1973; Nutter *et al.*, 1979). Results of these studies indicate that a wide variety of tree crops can be grown successfully in conjunction with land treatment of wastewater. Yields of some crops will be increased substantially, but special cultural procedures will be required for establishing and growing trees on irrigated sites. These studies in Michigan and in other states identify some species and hybrids that are particularly well adapted for wastewater application sites. They also identify species and stand conditions that should be avoided.

Table I

Percentage Difference in Survival and Height Growth Between Planted Trees Growing on Irrigated and Nonirrigated Sample Plots With Sand Soil in Michigan's Lower Peninsula.

Species/Hybrid	*Age*	*Survival*	*Height*
	(Yr)	----*(Percent)*----	
Populus x *euramericana*	5	0	+65*
Populus x *euramericana*	4	+46*	+169*
P. canescens x *p. grandidentata*	5	+24	+34
P. canescens x *p. tremuloides*	3	-17	+97
Green ash	5	+5	+47*
Tulip poplar	5	+39*	+56
European larch	5	+6	+76
Japanese larch	5	+2	+60
Northern white cedar	5	+10	+67*
Northern red oak	5	+11	+18
White spruce	4	+18	+118
Balsam fir	4	+44	+175
Douglas fir	4	-9	+50
Scotch pine varieties:			
Austrian hills	4	+3	+68
Scotch highland	4	+19	+78
French auvergne	4	+45	+83

*Difference between irrigated and nonirrigated trees is significant at $p \leq 0.10$.

This paper highlights research results that pertain to selecting and managing trees on effluent application sites. But a statement about irrigation rates should be made first.

IRRIGATION RATES

Because removal of pollutants is the primary objective of wastewater application to land, irrigation rates must be consistent with maintenance of water quality standards regardless of crop requirements. In Michigan and in many other states, that means that nitrate-nitrogen in the leachate cannot exceed 10 mg/l. In Pennsylvania, year-round irrigation with mechanical treatment plant effluent at the rate of 5.0

cm/wk increased nitrate-nitrogen in leachate to more than 10 mg/l. With growing season irrigation at 2.5 cm/wk, annual average concentrations of nitrate-nitrogen did not exceed 10 mg N/l during the first 6 years but did tend to increase (Sopper and Kerr, 1979).

In the short run, nitrate leaching is not likely to be as limiting with oxidation pond effluent as with mechanical treatment plant effluent because the mineralized forms of nitrogen in the oxidation pond effluent are lower during the growing season (King, 1978). Excessive loading of the soil with organic forms of nitrogen could ultimately result in prohibitively high nitrate leaching. It is also possible that phytotoxic constituents of wastewater, such as boron, might govern irrigation rates. If rates are not limited by either nitrate leaching or phytotoxic substances, efficient use of water to attain production objectives should be considered.

WOOD, FIBER, AND FUEL PRODUCTION

Several species and hybrids of *Populus* seem ideally suited for effluent irrigation sites. Irrigation can enhance establishment, especially with dormant cuttings, in marginal soils and during dry seasons. Growth can be dramatic. On a loamy sand soil in west-central lower Michigan, irrigation increased 4 year height growth of a *Populus* x *euramericana* hybrid by 169 percent and increased dry weight production, including foliage, by 1,200 percent. This hybrid produced more than twice as much biomass as a *Populus canescens* x *p. grandidentata* hybrid, the next best selection in the test. Nitrogen uptake in the fastest growing hybrid was comparable to uptake reported for corn and several forage grasses (Cooley, 1978). On a sand soil in northwestern lower Michigan, 3 year height growth of a *Populus canescens* x *p. tremuloides* hybrid was nearly doubled by irrigation (Cooley, 1979).

Other trees that have responded favorably to irrigation during their seedling stage include green *(Fraxinus pennsylvanica)* and white ash *(F. americana)*, tulip poplar *(Liriodendron tulipifera)*, white *(Picea glauca)* and Norway spruce *(P. abies)*, northern white-cedar *(Thuja occidentalis)*, European *(Larix decidua)* and Japanese larch *(L. leptolepis)*, white *(Pinus strobus)*, red *(P. resinosa)*, pitch *(P. rigida)* and Austrian pine *(P. nigra)*, Lombardy poplar *(Populus nigra* var. *italica*), and Douglas fir *(Pseudotsuga menziesii)*(Cooley, 1979; Sopper and Kardos, 1973; Neary *et al.*, 1975; Cole and Schiess, 1978). Lombardy poplar in Washington and *Populus* x *euramericana* hybrid in Michigan have far exceeded all others in production.

Existing stands have responded variably to irrigation. In experiments at Pennsylvania State University, 9 year height growth of white spruce was increased 360 percent by irrigation and diameter growth was increased 122 percent. Diameter growth of irrigated 30 to 50 year old red maple *(Acer rubrum)* and sugar maple *(A. saccharum)* was more than 300 percent greater than that of nonirrigated maple. In the same stand, irrigation increased diameter growth of trembling aspen *(Populus tremuloides)* by nearly 100 percent and mixed oaks *(Quercus alba* L., *Q. prinus* L., *Q. velutina* L., *Q. rubra* L., and *Q. coccinea* Muenchh*)* by as much as 69 percent (Sopper and Kardos, 1973). Dosages of 2.5 and 5.0 cm/wk of secondary treatment plant effluent were applied to this stand for varying times, ranging from an 8 week growing season to the whole year.

In a Southern Appalachian (Georgia) pine-oak stand, 3 years of irrigation increased average diameter growth by 29 percent and average height growth by 22 percent. The pine component in this stand consisted of white pine and Virginia pine *(P. virginiana* Mill*)*. The average tree diameter was 133 cm and the average height 11.2 m. Soils were sandy loam over clay and clay loam. The stand was irrigated with 7.6 cm/wk of oxidation pond effluent (Nutter *et al.*, 1979).

In a 50 year old beech-maple stand growing in loamy sand to sand soils in northwestern lower Michigan, effluent irrigation at rates of 3.8 and 7.6 cm/wk is increasing available nutrients, especially nitrogen, by adding them with the water, by accelerating decomposition of organic matter, and by enhancing the mycorrhizae that facilitate plant uptake (Otto, 1980). Phosphorus, calcium, and magnesium have increased significantly in the humus and upper mineral soil layers, and base saturation of the exchange complex has increased by 300 percent. The pH has increased from 4.5 to 6.8 in the humus and from 4.0 to 6.3 in the upper soil.

After 5 years, effluent irrigation's only measurable effect on growth has been to accelerate height growth of seedling size beech *(Fagus grandifolia)*. However, it has raised nitrogen concentrations in foliage of overstory beech, seedling-size beech, sugar maple, and red maple, and in all herbaceous plants (Table II). Even though growth of beech reproduction has been affected more than growth of other species in the stand, irrigation is favoring establishment and survival of species other than beech. Beech now makes up less of the reproduction on irrigated plots than on nonirrigated plots (Table III).

Pole-size red pine has not responded well to irrigation. At Pennsylvania State where trees were growing on loam with clay subsoil, irrigation with 5 cm of effluent per week reduced height growth. Irrigated trees were completely destroyed by snow and wind after 6 years of irrigation (Sopper and Kardos, 1973). In lower Michigan on a loamy sand soil with sand and gravel subsoil, no windthrow has occurred, but irrigation with up to 8.8 cm of effluent per week has failed to elicit any increase in diameter or height growth (Cooley, 1979). In addition, the trees have exhibited boron toxicity symptoms (needle tip necrosis) since the second year of irrigation (Neary *et al.*, 1975). Lack of response may be due to increases in soil pH, from 4.9 or less to 6.5 or more (Harris, 1979) which is outside the optimal range for red pine (Rudolf, 1957). On the other hand, the rapid growth without irrigation may represent the maximum inherent growth potential for the trees planted on this site.

Table II

Effect of Irrigation for 4 Years on Total Kjeldahl Nitrogen (TKN) Content of Foliage in a Beech-Maple Stand in Northwestern Lower Michigan.

	Irrigation Rate (cm/wk)		
Species	*0*	*3.8*	*7.6*
	(Percent of Dry Weight)		
	Overstory		
Sugar Maple	1.353a*	1.541a	1.417a
Red Maple**	1.704	1.548	1.814
Beech	1.677a	1.885b	2.130c
	Reproduction		
Sugar Maple	2.057a	2.392b	2.685c
Red Maple	1.931a	2.241ab	2.729b
Beech	2.429a	2.519a	2.893b
	Spring Ephemerals		
All	1.887a	2.847b	2.748b

*Within species, values followed by different letters are statistically different (p = 0.10).

**Not enough samples for statistical testing.

CHRISTMAS TREE PRODUCTION

Some sewage treatment authorities may prefer to raise Christmas trees rather than wood products. Christmas trees can be produced in a shorter rotation, require less specialized equipment, and can be marketed in smaller quantities. They are one of the few tree crops that can be grown under moving overhead sprinkler systems such as the center pivot irrigator.

Table III

Species Composition of Established Reproduction in a Beech-Maple Stand in Northwestern Lower Michigan Before and After 4 Years of Irrigation.*

	Irrigation Rate (cm/wk)					
	0		*3.8*		*7.6*	
Species	*1975*	*1979*	*1975*	*1979*	*1975*	*1979*
	Not Thinned (Percent of Total)					
Beech	81.4	64.4	87.3	32.8	53.9	31.3
Sugar Maple	17.9	28.6	11.3	64.1	46.1	60.4
Other Species	0.7	7.5	1.4	3.1	0.0	8.4
	Thinned					
Beech	60.6	44.0	48.1	26.8	47.4	21.7
Sugar Maple	32.4	41.5	45.4	47.0	25.9	52.4
Other Species	7.0	14.5	6.4	26.2	26.7	5.5

*Stems 15 to 299 cm tall.

On a test site in northwestern lower Michigan, Scotch pine *(Pinus sylvestris)*, white spruce, and balsam fir *(Abies balsamea)* have all responded well to effluent irrigation. The test site is an abandoned farm field and pasture with well-drained sand soil. Without irrigation, Scotch pine is the only one of the species tested that could be grown successfully, but with regular applications of effluent, white spruce and balsam fir could also be grown. Irrigation increased balsam fir survival by 44 percent and its growth after establishment by 850 percent. White spruce survival was increased by 18 percent and its growth after establishment by 118 percent.

Because of inherent differences and differences in stock characteristics, response varied among the three Scotch pine varieties that were tested. Increases in both survival and growth were directly proportional to the shoot:root ratios of the different varieties. Irrigation increased survival of the variety with the highest shoot:root ratio (French auvergne) by 45 percent but had little effect (3%) on survival of the variety with the lowest shoot:root ratio (Austrian hills). It increased height growth of the French auvergne variety by 197 percent but Austrian hills by only 93 percent.

Even with irrigation, the Douglas fir used was not suitable for Christmas tree production on this site because it was damaged repeatedly by frost. Its survival rate was poorer where it was irrigated. Even though irrigated trees were 59 percent taller than control trees, they were only about 30.5 cm tall after 4 years (Cooley, 1980).

SPECIAL PROBLEMS

Effluent irrigation has several side effects that can cancel all its benefits. Foremost, it enhances the growth of grass and other herbaceous plants. In irrigated abandoned fields, existing cover can smother most planted seedlings before they become established enough to benefit from irrigation. After trees are established, grass and weeds provide cover for mice and rabbits which can cause extensive damage. One irrigated plantation of hybrid poplars in lower Michigan was virtually destroyed by mice even though they did little damage to the same hybrids on an adjacent control plot (Cooley, 1979). Mice also destroyed nearly a third of the irrigated Christmas trees in a nearby plantation.

Clean cultivation would, of course, preclude weed competition and rodent damage, but it could also lead to serious erosion on any but the flattest sites. Furthermore, vegetative cover serves as a sink for nutrients, especially nitrogen, while the trees are becoming established. With pond effluents, such a sink may not be critical because much of the nitrogen is in organic form. However, effluent from mechanical treatment has a higher concentration of ammonium and nitrate-nitrogen, and the herbaceous cover may be necessary to prevent excessive nitrate leaching. Therefore, vegetative cover should be managed so that it does not smother planted seedlings nor provide cover for rodents. Establishing trees in old fields might require a combination of spraying with herbicides along the tree rows until the trees are established and mowing between rows to keep vegetation short.

In selecting herbicides for weed control, the effect of heavy and frequent irrigation on tree root development and herbicide movement must be taken into account. These factors limit the use of simazine and other chemicals that will injure trees if they are leached out of the surface soil. Directed sprays of foliage-applied herbicides are less likely to cause damage. During the early stages of establishment, weed control measures will have to be applied frequently, because even though existing weeds are eliminated, frequent irrigation will encourage regrowth.

Shallow root development, a consequence of frequent irrigation, will make trees extremely susceptible to drought injury. One Christmas tree planting was not irrigated during May, June, or July of the fifth growing season because the irrigator was broken down. This stoppage killed 41 percent of the white spruce, 72 percent of the balsam fir, and 38 percent of the Douglas fir. Very few of the Scotch pine in the plantation died, but shoot growth was so short that sheared trees did not develop the crown density required for Christmas trees. This experience illustrates the need to avoid any extended interruptions in the irrigation schedule once trees are adapted to frequent irrigation.

Even though wastewater contains some nutrients, nutrient deficiences may still occur because growth is not limited by lack of moisture. For example, where *Populus* x *euramericana* hybrids were irrigated with 3.0 cm/wk, nitrogen uptake exceeded input in sewage lagoon wastewater during the fourth year after planting. Where they were irrigated with 7.0 cm/wk, uptake approached input by the fifth year. It may be necessary to supplement wastewater with chemical fertilizers to maintain maximum growth of some tree crops even though sewage lagoons are the source of irrigation.

CONCLUSIONS

A wide variety of species and *Populus* hybrids are compatible with slow rate application of sewage effluent on sand soils in the Great Lakes region. The most appropriate tree crop will depend largely on the importance of plant uptake to the treatment process, availability of markets for the products grown, and the level of management available. Generally speaking, irrigation is more likely to increase growth in young trees, either planted or natural, than in pole-size or larger trees. Because growth rate is a major determinant of nutrient uptake and assimilation, young trees will contribute more than old trees to the treatment.

Populus hybrids seem especially well adapted for fiber production and nutrient uptake on effluent application sites. The few clones tested have grown much better with irrigation than without it on well-drained sand soils. Because insects and diseases can severely damage hybrid poplars, resistant clones must be selected. However, information about insect and disease resistance is now available for only a few clones on a limited range of site conditions. Without more precise information, several clones that have grown well on sites similar to the effluent application site should each be planted in blocks of 1 to 1.5 hectares. Those clones that are then severely damaged by insects or diseases should be replaced as soon as the damage becomes apparent.

Effluent can also be used to increase growth rate in stands of naturally occurring regeneration or plantations of indigenous species. If openings are maintained around the sprinklers, irrigation with fixed sets can continue at least until the stands reach pole size. Irrigation may not increase growth of trees beyond sapling size, but it will not reduce growth or cause excessive mortality in the deep, well-drained sand soils of the northern Great Lakes region.

Christmas trees are the only tree crop for which overhead sprinkler systems can be utilized throughout the rotation. Scotch pine, white spruce, and balsam fir can be grown successfully even on sites that are too droughty to grow spruce or fir without irrigation.

Weed control is essential for establishing hybrid poplar or conifer plantations on sites irrigated with effluent. To preclude excessive damage by rodents, weed control must continue until rank growth of herbaceous plants is reduced by shade from planted trees or until tree bark is too thick for mice to gnaw through. Great care must be taken when using herbicides that can injure trees when leached to the tree's rooting zone. Mowing and directed sprays of foliage-applied herbicides are less likely to damage trees.

If irrigated trees show symptoms of nutrient deficiencies, supplemental mineral nutrients should be added to the irrigation water or directly to the site.

LITERATURE CITED

Cole, D.W. and P. Schiess. 1978. Renovation of Wastewater and Response of Forest Ecosystems: The Pack Forest Study. In H.L. McKim, Proc. State of Knowledge in Land Treatment of Wastewater. U.S. Army Cold Regions Research and Engineering Laboratory, Hanover, NH 03755, August 20-25, 2:323-331.

Cooley, J.H. 1980. Christmas Trees Enhanced by Sewage Effluent. Compost Science/Land Utilization 21(6):28-30.

Cooley, J.H. 1979. Effects of Irrigation With Oxidation Pond Effluent on Tree Establishment and Growth on Sand Soils. In W.E. Sopper and S.N. Kerr (eds.), Utilization of Municipal Sewage Effluent and Sludge on Forest and Disturbed Land. The Pennsylvania State University Press, University Park, PA 16802, pp. 145-153.

Cooley, J.H. 1978. Nutrient Assimilation in Trees Irrigated With Sewage Oxidation Pond Effluent. In P.E. Pope (ed.), Proc. Central Hardwoods Forest Conference II, Purdue University, West Lafayette, IN 47907, pp. 323-340.

Harris, A.R. 1979. Physical and Chemical Changes in Forested Michigan Sand Soils Fertilized With Effluent and Sludge. In W.E. Sopper and S.N. Kerr (eds.), Utilization of Municipal Sewage Effluent and Sludge on Forest and Disturbed Land. The Pennsylvania State University Press, University Park, PA 16802, pp. 155-161.

King, D.L. 1978. The Role of Ponds in Land Treatment of Wastewater. In H.L. McKim (ed.), Proc. State of Knowledge in Land Treatment of Wastewater. U.S. Army Corps of Engineers Cold Regions Research and Engineering Laboratory, Hanover, NH 03755, August 20-25, 2:191-198.

Little, S., H.W. Lull, and I. Remson. 1959. Changes in Woodland Vegetation and Soils After Spraying Large Amounts of Wastewater. Forest Sci. 5:18-27.

Neary, D.G., G. Schneider, and D.P. White. 1975. Boron Toxicity in Red Pine Following Municipal Wastewater Irrigation. Soil Sci. Soc. Amer. Proc. 39:981-982.

Nutter, W.L., R.C. Schultz, and G.H. Brister. 1979. Renovation of Municipal Wastewater by Spray Irrigation on Steep Forest Slopes in the Southern Appalachians. In W.E. Sopper and S.N. Kerr (eds.), Utilization of Municipal Sewage Effluent and Sludge on Forest and Disturbed Land. The Pennsylvania State University Press, University Park, PA 16802, pp. 77-85.

Otto, P.C. 1980. The Effects of Sewage Effluent on *Acer* sp. Mycorrhizae and Related Soil Properties. M.S. Thesis, University of Michigan, Ann Arbor, MI 48109, 65 pp.

Rudolf, P.O. 1957. Silvical Characteristics of Red Pine *(Pinus resinosa)*. USDA, Forest Service Lake States Experiment Station, Station Paper 44, St. Paul, MN 55455, 32 pp.

Rudolph, V.J. 1957. Further Observations on Irrigating Trees With Cannery Wastewater. Mich. Agr. Exp. Sta. Quart. Bull. 39:416-423.

Rudolph, V.J. and R.E. Dils. 1955. Irrigating Trees With Cannery Wastewater. Mich. Agr. Exp. Sta. Bull. 37:407-411.

Sopper, W.E. and L.T. Kardos. 1973. Vegetation Responses to Irrigation With Treated Municipal Wastewater. In W.E. Sopper and L.T. Kardos (eds.), Recycling Treated Municipal Wastewater and Sludge Through Forest and Cropland. The Pennsylvania State University Press, University Park, PA 16802, pp. 271-294.

Sopper, W.E. and S.N. Kerr. 1979. Renovation of Municipal Wastewater in Eastern Forest Ecosystems. In W.E. Sopper and S.N. Kerr (eds.), Municipal Sewage Effluent and Sludge on Forest and Disturbed Land. The Pennsylvania State University Press, University Park, PA 16802, pp. 61-76.

CHAPTER 10

TREE SEEDLING RESPONSES TO WASTEWATER IRRIGATION ON A REFORESTED OLD FIELD IN SOUTHERN MICHIGAN

Dale G. Brockway
Water Quality Division
Michigan Department of Natural Resources
Lansing, Michigan 48917

INTRODUCTION

In recent years, the number of municipal wastewater irrigation projects utilizing forest ecosystems for nutrient recycling and groundwater recharge has continued to grow. Land managers involved with these projects have, in the same time period, increased their demand for scientific data which better defines the capabilities and limitations of various forest species-site combinations. Smith and Evans (1977) suggested that an initial step in planning for wastewater recycling in forests was to assess the species ecological requirements in terms of water and nutrients. Identifying forest species adapted to the moisture and nutrient regimen provided by wastewater irrigation upon a specific type of forest site is a prerequisite to sound irrigation project management.

Old fields have come under the increased scrutiny of land managers as potential forest sites where plantations established near municipalities could benefit from wastewater irrigation. Old fields occupy millions of hectares in the eastern United States and are typically the target of reforestation efforts by private and public land managers. These sites have a history of agricultural management which has significantly modified their native vegetation and soil characteristics. Old fields contain a well developed Ap soil horizon supporting a diverse array of native and exotic herbaceous and woody plant species. Intense competition from these species for available water and nutrients is a major obstacle to tree plantation establishment and development. Cultural treatments such as wastewater irrigation, which supplement site water and nutrient resources, may enhance seedling survival and growth if weed competition is controlled. Weed control is particularly critical for hardwood seedlings, which are often less able than conifers to compete with established grasses and associated vegetation.

In 1974, three conifer and six hardwood tree species were planted on an old field site and irrigated with municipal wastewater. The principal study objective was by monitoring seedling survival, growth and nutrient status, to assess tree species suitability for plantation establishment on old field sites under wastewater irrigation. A secondary objective was to determine the wastewater renovation capacity of the reforested old field.

MATERIALS AND METHODS

Detailed establishment procedures and site description for the 2.1 ha old field were reported earlier (Brockway *et al.*, 1979a). The study area was located 5 km south of the main campus of Michigan State University at East Lansing where gently rolling glacial till, 20 m in depth, overlies a bedrock of Saginaw sandstone. Prior to cultivation, the area supported a beech-maple forest. During the post-cultivation period, a well developed sod and herbaceous layer, including *Solidago*, *Digitaria*, *Agropyron* and *Andropogon* developed on the site. The Miami-Conover-Brookston Alfisol soil catena dominated the site. Annual precipitation averaged 765 mm.

The site was planted in April, 1974, following a pre-planting weed control treatment of paraquat-CL (1.1 kg/ha) and simazine (2.2 kg/ha). The nine tree species planted were black cherry *(Prunus serotina* Ehrh.*)*, black walnut *(Juglans nigra* L.*)*, eastern cottonwood *(Populus deltoides* Bartr.*)*, northern red oak *(Quercus rubra* L.*)*, white ash *(Fraxinus americana* L.*)*, tulip poplar *(Liriodendron tulipifera* L.*)*, Scotch pine *(Pinus sylvestris* L.*)*, Norway spruce *(Picea abies* Karst.*)*, and white spruce *(Picea glauca* Voss.*)*. In the first four years following planting, tree rows were annually weeded using glyphosate (1.4 kg/ha) and weeds between rows were mowed semi-annually. Weeds along irrigation lines were controlled with an annual simazine-atrazine treatment. Since 1976, the trees have been side pruned and basal sprouts removed to encourage single stem development.

The experimental design consisted of 14 randomized complete blocks. Each block contained 9 tree species organized in 9 rows, each 63 m long and containing 40 individual trees of a single species, totaling 5040 seedlings in the entire plantation. Seedling spacing was 1.5 m x 2.1 m with each block delineated by intervening irrigation lines. Block 1 was the control, receiving no irrigation, while blocks 2 through 14 were treated twice weekly with municipal wastewater.

Secondarily treated wastewater obtained from the East Lansing Municipal Wastewater Treatment Plant was cycled at the Michigan State University Water Quality Management Facility through a system of four ponds ranging in size from 3 to 5 ha. Water directly from the East Lansing plant plus water back-siphoned from the first pond was delivered to the site by an overhead spray irrigation at a rate of 4 mm/hr, totaling 51 mm/wk for approximately 15 weeks each season over seven years. Average annual nutrient application rates with irrigation were 160 kg total nitrogen, 30 kg phosphorus, and 130 kg potassium per hectare. Nitrogen fraction application rates average 90 kg nitrate and 20 kg ammonia per hectare. Application rates of micronutrients were very low, generally less than 1 kg/ha.

During the initial four years of this study, nutrient cycling dynamics received great attention. Weekly water samples were obtained from sprayheads and suction lysimeters installed to sample soil water at a depth of 61 cm in the soil profile. Water samples were preserved with concentrated sulfuric acid (1 ml/600 ml) or concentrated nitric acid (1 ml/300 ml), stored at 4°C, and analyzed by the Institute of Water Research Laboratory at Michigan State University and the University of Wisconsin. Soil samples were collected following each irrigation season at depths of 0-15, 15-30, 45-60, and 105-120 cm in regularly spaced intervals across the site. Soil samples were air dried, composited by soil series and depth, pulverized and passed through a 2 mm screen. Nutrient analysis was performed at the Michigan State University Soil Chemistry Laboratory and A & L Agricultural Laboratories of Fort Wayne, Indiana. Measurement of seedling nutrient status was conducted each September by collecting foliage samples from all trees in blocks 1 through 10. Foliage samples were composited by row, oven dried at 75°C, ground in a Wiley mill, and passed through a 20 mesh screen. Nutrient concentrations were determined at the Michigan State University Plant Tissue Analysis Laboratory.

Tree seedling growth was measured in the initial four years of this study by destructive methods, requiring total above ground tree harvest to obtain biomass data. One seedling per row was selected, cut at the groundline, oven dried, and weighed to determine biomass of the total tree, stem, and foliage components. During each year of measurement, tree height, basal stem diameter (15 cm above ground), and where possible, diameter at breast height (1.4 m above ground) were recorded in the field.

RESULTS AND DISCUSSION

Seedling Survival and Growth

Tree seedling survival was not enhanced by wastewater irrigation (Table I). Following seven seasons of irrigation, modest survival increases in eastern cottonwood, white ash, and white spruce were noted. Decreased survival of treated seedlings occurred in Norway spruce, black cherry, black walnut, and especially tulip poplar and red oak. The severely reduced survival rates of the latter two species were a result of excessive weed competition. Where weeds were not adequately controlled, irrigation treatment stimulated their growth to the point of physically overwhelming the tree seedlings. Scotch pine survival was unaffected by irrigation. Overall survival rates of the above were similar to those reported by Cooley (1979) for species of the same genera.

Table I

Species Survival After 7 Seasons of Wastewater Irrigation, 1980.

Species	Seedling Survival	
	Control	Irrigated
	-----------%-----------	
Eastern Cottonwood	70	81
White Ash	97	100
Scotch Pine	97	97
Tulip Poplar	87	57
Black Walnut	78	73
Norway Spruce	100	97
White Spruce	78	97
Black Cherry	95	87
Red Oak	87	54

Table II shows the cumulative growth of plantation trees from 1974 to 1980. Irrigation in that period produced significant increases in height and diameter attainment for eastern cottonwood, white ash, Scotch pine, tulip poplar, and black walnut. Norway spruce and white spruce height growth were unaffected by irrigation. Black cherry and red oak height and diameter growth were found to decrease with irrigation treatment.

Table II

Species Growth After 7 Seasons of Wastewater Irrigation, 1980.

Species	*Height*		*Diameter at Breast Height*		*Basal Stem Diameter*	
	Control	*Irrigated*	*Control*	*Irrigated*	*Control*	*Irrigated*
	----------*m*----------		----------*cm*----------			
Eastern Cottonwood	2.21	4.42*	2.3	6.4*	4.1	9.1
White Ash	1.47	2.49*	1.3	2.8*	2.0	3.8*
Scotch Pine	2.03	2.36*	3.3	4.1*	---	---
Tulip Poplar	0.69	2.11*	-0-	2.5*	1.0	4.1*
Black Walnut	1.22	1.45*	-0-	1.8*	2.5	3.3*
Norway Spruce	1.35	1.35	-0-	-0-	---	---
White Spruce	1.04	0.94	-0-	-0-	---	---
Black Cherry	1.50	1.22	1.0	-0-	2.3	1.5
Red Oak	1.75	1.14	1.8	-0-	2.8	1.5

*Significantly greater than control at the 0.05 level.

Wastewater irrigation has been cited in several studies as a silvicultural treatment which ameliorates site conditions for plant growth (Smith and Evans, 1977). Einspahr *et al.* (1972) have demonstrated that in some woody plants, height growth is stimulated primarily by applied water and diameter growth is increased by added nutrients. The tree growth results of this study paralleled those of Cooley (1979), Settergren *et al.* (1974), Sutherland *et al.* (1974), and Post and Smith (unpublished)*. These researchers found that the optimum growth response to wastewater irrigation was exhibited by lowland hardwood species of the genera *Populus* and *Fraxinus*. Moderate growth responses were seen in mesic-site upland hardwoods and poor responses were measured in dry-site pines and oaks. The growth results in this study were not consistent with those of Sopper and Kardos (1973) who reported significant growth increases for oak and white spruce stands irrigated with wastewater. The disparity in results could be ascribed only to the fact that their studies utilized older, well established forest stands while this study examined the growth of newly planted seedlings whose root systems incompletely occupied the old field site.

Irrigation extended the season during which seedling growth could occur. Flushes of growth observed on this plantation extended much later into the summer season for irrigated trees than for controls, presumably the result of favorable soil moisture conditions. Similar findings were reported by Howe (1968) for irrigated ponderosa pine and by Kaufman (1968) for white pine and loblolly pine. When exposed to water stress, these pine seedlings underwent root suberization which precipitated the onset of dormancy. Where water stress was severe, the intercalary regions along each shoot also became dormant.

Seedling Nutrient Concentrations and Total Assimilation

After four years of wastewater irrigation, foliar nutrient levels ranged from low to intermediate, within the tolerable limits reported for these species (Bengtson *et al.*, 1968). Foliar nitrogen concentrations were within the higher portion of that range, however, exceeding 2 percent in most species and reaching 3 percent or more in two of the species (Table III). Irrigated trees generally contained significantly higher nutrient concentrations than unirrigated controls, an indication that the irrigated trees had assimilated nutrients applied in the wastewater. Nutrient dilution

---

**D.M. Post and W.H. Smith, unpublished report, School of Forest Resources and Conservation, University of Florida, Gainsville, FL 32601.*

resulting from increased growth in the irrigated group may have accounted for the lack of statistical difference with controls in the concentrations of certain nutrients. Unlike the results reported by Neary *et al.* (1975), no nutrient toxicity or deficiency symptoms were noted among plantation trees.

Leaf biomass and leaf nutrient concentrations, as an index of total tree nutrient assimilation following the 1977 season, are shown in Table IV. Eastern cottonwood and Scotch pine accumulated the greatest nutrient quantities in their foliage followed by white ash. These species would probably concentrate large nutrient reserves in their harvestable woody parts as well. As Kramer and Kozlowski (1960) pointed out, as much as 50 percent of the foliar content of some nutrients is retranslocated into the stem prior to leaf abscission. The remaining species assimilated lesser total nutrients as a result of slower growth rates and a resulting lower biomass.

Seedling Performance and Species Recommendations

The tree species evaluated in this study for wastewater irrigation suitability on old field sites were arranged on an ordinal scale shown in Table V. The performance of each species was based largely upon growth response to irrigation; however, total nutrient assimilation and survival were also considered in the final recommendation.

Eastern cottonwood was found to be "superior" in overall performance to all other species tested and as such is highly recommended for use in wastewater irrigation projects. This result was not surprising, when the hydrophilic nature of this species is considered. White ash, Scotch pine, tulip poplar, and black walnut exhibited "good" performance under wastewater irrigation and are also recommended for wastewater recycling programs. It was noted, however, that tulip poplar may suffer extensive mortality on irrigated old field sites unless weed control is adequate. Norway spruce and white spruce were found to be "marginal" in their performance after seven years of irrigation. Growth benefits on the old field were not discernible and overall nutrient assimilation was portionally diminished. Other forest sites may produce more encouraging results with these species than do old fields. Black cherry and northern red oak exhibited decreased survival and growth and very low total nutrient assimilation with wastewater irrigation. Whether their "poor" performance was a result of weed competition or other biotic interactions, these species are not recommended for reforestation of old fields irrigated with municipal wastewater.

Table III

Foliar Nutrient Concentrations by Species Following Wastewater Irrigation for the 1977 Growing Season (Brockway et al., 1979b).

Species	N		P		K		Na	
	C	I	C	I	C	I	C	I
	%							
Eastern Cottonwood	1.9	1.9	0.18	0.24*	0.48	0.42	0.05	0.19*
Scotch Pine	2.7	3.0*	0.30	0.32	0.82	0.86*	0.05	0.11*
White Ash	2.5	2.5	0.25	0.26	0.52	0.97*	0.05	0.07
Norway Spruce	2.0	2.6*	0.40	0.39	1.00	1.23*	0.06	0.13*
Black Walnut	2.0	2.2*	0.19	0.19	0.66	0.52	0.08	0.23*
Tulip Poplar	1.8	1.8	0.05	0.08*	0.66	0.65	0.06	0.13*
Black Cherry	2.5	2.5	0.20	0.19	0.62	0.57	0.04	0.09
Red Oak	2.0	2.0*	0.32	0.22	0.68	1.03*	0.04	0.16
White Spruce	2.4	3.2*	0.40	0.26	0.88	1.37*	0.07	0.36*

Species	Ca		Mg		B		Mn		Zn	
	C	I	C	I	C	I	C	I	C	I
	%				mg/kg					
Eastern Cottonwood	0.78	1.12*	0.14	0.16	13.7	42.1*	172	185	35	36*
Scotch Pine	2.20	1.36	0.38	0.40	17.4	21.1*	279	174	35	29
White Ash	0.71	1.11*	0.38	0.36	22.0	40.1*	136	92	13	18
Norway Spruce	0.93	0.91	0.40	0.41	53.6	54.4	157	96	35	37
Black Walnut	0.93	1.15*	0.13	0.18*	28.4	54.2*	386	180	26	13
Tulip Poplar	0.61	0.81*	0.15	0.12	22.0	35.2	107	140	24	6
Black Cherry	0.66	0.55	0.23	0.32*	46.1	45.6	253	367	11	11
Red Oak	0.81	1.18*	0.30	0.23	22.0	25.9*	38	29	40	27
White Spruce	0.86	1.27*	0.50	0.39	39.6	42.8	52	71	51	35

* = Significantly greater than control at the 0.05 level.

C = Control

I = Irrigated

Table IV

Foliar Nutrient Concentration and Content by Species After 4 Years of Wastewater Irrigation, 1974-1977 (Brockway et al., 1979b).

Species	*Leaf Biomass*	*Concentration*							
		N		*P*		*K*		*Na*	
	*gpt**	*%*	*gpt*	*%*	*gpt*	*%*	*gpt*	*%*	*gpt*
Eastern Cottonwood	940	1.3	17.9	0.24	2.3	0.42	3.9	0.19	1.8
Scotch Pine	441	3.0	13.2	0.32	1.4	0.86	3.8	0.11	0.5
White Ash	149	2.5	3.7	0.26	0.4	0.97	1.4	0.07	0.1
Norway Spruce	127	2.6	3.3	0.39	0.5	1.23	1.6	0.13	0.2
Black Walnut	126	2.2	2.8	0.19	0.2	0.52	0.6	0.23	0.3
Tulip Poplar	136	1.8	2.5	0.08	0.1	0.65	0.9	0.13	0.2
Black Cherry	75	2.5	1.9	0.19	0.1	0.57	0.4	0.09	0.1
Red Oak	93	2.0	1.9	0.22	0.2	1.03	1.0	0.16	0.1
White Spruce	48	3.2	1.5	0.26	0.1	1.37	0.7	0.36	0.2

Species	*Concentration*									
	Ca		*Mg*		*B*		*Mn*		*Zn*	
	%	*gpt*	%	*gpt*	*mg/kg*	*mg PT***	*mg/kg*	*mg PT*	*mg/kg*	*mg PT*
Eastern Cottonwood	1.12	10.5	0.16	1.5	42.1	39.6	185	174.9	36	33.8
Scotch Pine	1.36	6.0	0.40	1.8	21.1	9.3	174	76.7	29	12.8
White Ash	1.11	1.7	0.36	0.5	40.1	6.0	92	13.7	18	2.7
Norway Spruce	0.91	1.2	0.41	0.5	54.4	6.9	96	12.2	37	4.7
Black Walnut	1.15	1.5	0.18	0.2	54.2	6.8	180	22.7	13	1.6
Tulip Poplar	0.81	1.1	0.12	0.2	35.2	4.8	140	19.1	6	0.8
Black Cherry	0.55	0.4	0.32	0.2	45.6	3.4	367	27.5	11	0.8
Red Oak	1.18	1.1	0.23	0.2	25.9	2.4	29	2.7	27	2.5
White Spruce	1.27	0.6	0.39	0.2	42.8	2.0	71	3.4	35	1.7

* = grams per tree

** = milligrams per tree

Table V

Species Performance Summary: Increase or Decrease in Growth With Respect to Control Species Over 7 Years.

Species	*Height*	*Diameter at Breast Height*	*Basal Stem Diameter*	*Performance Class*
	%			
Eastern Cottonwood	+100	+178	+122	Superior
White Ash	+69	+115	+90	Good
Scotch Pine	+16	+24	---	Good
Tulip Poplar	+206	---	+310	Good
Black Walnut	+19	---	+32	Good
Norway Spruce	0	---	---	Marginal
White Spruce	-10	---	---	Marginal
Black Cherry	-19	---	-35	Poor
Red Oak	-35	---	-46	Poor

Wastewater Renovation Capacity of the Site

The old field study site provided adequate water renovation for the most troublesome of nutrients, nitrate-nitrogen, and phosphorus. In 1977, nitrate and total nitrogen removal from the applied wastewater approximated 85 percent by the time it had percolated to a depth of 61 cm in the soil profile. This value exceeded that reported by Neary (1974) on similar soils. Phosphorus removal was excellent, exceeding 98 percent.

The values reported here for nitrate and phosphorus were similar to those reported by Leland *et al.* (1978) for a nonforested old field site. This comparison underscores the importance of grasses and associated herbs in renovating land-applied wastewater on old field sites. During the course of this study, the plantation seedlings did not fully occupy the site with their immature root systems and incompletely formed crowns. Although evidence on water and nutrient uptake was obtained, previously established grasses and associated vegetation and soil adsorption were thought responsible for a greater amount of wastewater renovation. As the seedling root systems develop and more fully occupy the site, an increasing proportion of the wastewater renovation in this ecosystem may in the future be attributed to the trees.

SUMMARY

During the seven years following establishment, a conifer-hardwood plantation planted on an old field site in southern Michigan was irrigated with 51 mm of municipal wastewater per week. The suitability of the planted tree species for use in reforesting old field sites under wastewater irrigation was evaluated based upon growth, survival, and nutrient assimilation performance. Eastern cottonwood was found to exhibit superior growth and nutrient assimilation responses to irrigation and is highly recommended for this cultural use. White ash, Scotch pine, tulip poplar, and black walnut exhibited good performance under irrigation and are also recommended. Norway spruce and white spruce were found to be marginal in the above performance categories and while not recommended for use on irrigated old field sites, may prove useable on other forest sites under irrigation. Black cherry and northern red oak exhibited survival and growth rates negatively related to irrigation treatment and are not recommended for reforestation of old field sites irrigated with municipal wastewater.

The old field study site provided adequate water renovation, removal of nitrate-nitrogen, and total nitrogen approximating

85 percent and removal of phosphorus exceeding 98 percent. These values were similar to those reported for an unforested old field, underscoring the importance of grasses and associated herbs in renovating land-applied wastewater where planted seedlings have not yet fully occupied the site. As seedling root systems develop toward more complete site occupancy, a greater proportion of wastewater renovation may be attributed to the trees.

ACKNOWLEDGMENTS

This research project was supported by the McIntire-Stennis Cooperative Forestry Research Program (Project No. 3145) and the Office of Water Research and Technology (OWRT Project No. A-086-MICH) through the Michigan State University Institute of Water Research.

LITERATURE CITED

Bengtson, G.W., R.H. Brendemuehl, W.L. Pritchett, and W.H. Smith (eds.). 1968. Forest Fertilization. Tennessee Valley Authority, Muscle Shoals, AL 35660, 306 pp.

Brockway, D.G., G. Schneider, and D.P. White. 1979a. Dynamics of Municipal Wastewater Renovation in a Young Conifer-Hardwood Plantation in Michigan. In W.E. Sopper and S.N. Kerr (eds.), Utilization of Municipal Wastewater and Sludge on Forest and Disturbed Land. Pennsylvania State University Press, University Park, PA 16802, pp. 87-101.

Brockway, D.G., G. Schneider, and D.P. White. 1979b. Municipal Wastewater Renovation, Growth, and Nutrient Uptake in an Immature Conifer-Hardwood Plantation. In C.T. Youngberg (ed.), Forest Soils and Land Use. Fifth North American Forest Soils Conference Proceedings, Colorado State University, Fort Collins, CO 80521, pp. 565-584.

Cooley, J.H. 1979. Effects of Irrigation With Oxidation Pond Effluent on Tree Establishment and Growth in Sandy Soils. In W.E. Sopper and S.N. Kerr (eds.), Utilization of Municipal Wastewater and Sludge on Forest and Disturbed Land. Pennsylvania State University Press, University Park, PA 16802, pp. 145-153.

Einspahr, D.W., M.K. Benson, and M.L. Hardee. 1972. Influence of Irrigation and Fertilization on Growth and Wood Properties of Quaking Aspen. In Symposium Proceedings on Effects of Growth Acceleration on Wood. USDA, Madison, WI 53711, pp. I1-I8.

Howe, J.P. 1968. Influence of Irrigation on Ponderosa Pine. Forest Products J. 18:84-93.
Kaufman, M.R. 1968. Water Relations of Pine Seedlings in Relation to Root and Shoot Growth. Plant Physiology 43: 281-288.
Kramer, P.J. and T.T. Kozlowski. 1960. Physiology of Trees. McGraw-Hill Book Co., Inc., New York, NY 10020, 647 pp.
Leland, D.E., D.C. Wiggert, and T.M. Burton. 1978. Winter Spray Irrigation of Secondary Municipal Effluent in Michigan. In T.M. Burton and T.G. Bahr (eds.), Felton-Herron Creek, Mill Creek Pilot Watershed Studies. Institute of Water Research, Michigan State University, East Lansing, MI 48824, pp. 121-139.
Neary, D.G. 1974. Effects of Municipal Wastewater Irrigation on Forest Sites in Southern Michigan. Ph.D. Thesis, Michigan State University, East Lansing, MI 48824. University Microfilms, University of Michigan, Ann Arbor, MI 48106 (Diss. Abstr. 36-16B).
Neary, D.G., G. Schneider, and D.P. White. 1975. Boron Toxicity in Red Pine Following Municipal Wastewater Irrigation. Soil Sci. Soc. Amer. Proc. 30:981-982.
Settergren, C.D., J.A. Turner, and W.F. Hansen. 1974. The Use of Sewage Effluent Irrigation Techniques at Large Recreational Developments. In Proc. Society of American Forestry, Forest Issues in Urban America. Soc. Amer. For., Washington, DC, pp. 272-282
Smith, W.H. and J.O. Evans. 1977. Special Opportunities and Problems in Using Forest Soils for Organic Waste Application. In L.F. Elliott and J.J. Stevenson (eds.), Soils for Management of Organic Wastes and Waste Water. American Society of Agronomy, Madison, WI 53711, pp. 429-454.
Sopper, W.E. and L.T. Kardos. 1973. Vegetation Responses to Irrigation With Treated Municipal Wastewater. In W.E. Sopper and L.T. Kardos (eds.), Recycling Treated Municipal Wastewater and Sludge Through Forest and Cropland. Pennsylvania State University Press, University Park, PA 16802, pp. 271-293.
Sutherland, J.C., J.H. Cooley, D.G. Neary, and D.H. Urie. 1974. Irrigation of Trees and Crops With Sewage Stabilization Pond Effluent in Southern Michigan. In Proc. Wastewater Use in the Production of Food and Fiber. U.S. Environmental Protection Agency, EPA 660/2-74-041, Washington, DC 20460, pp. 295-313.

CHAPTER 11

STUDIES OF LAND APPLICATION IN OLD GROWTH FORESTS IN SOUTHERN MICHIGAN

Thomas M. Burton
Department of Zoology, Department of Fisheries and Wildlife, and Institute of Water Research
Michigan State University
East Lansing, Michigan 48824

INTRODUCTION

The use of late successional second-growth forests in southern Michigan for renovation and recharge of secondary municipal wastewater was studied on the Water Quality Management Facility (WMQF) at Michigan State University from 1976 through 1980. The objectives of these studies were (1) to ascertain if such forests were capable of removing enough phosphorus and nitrogen from wastewater to achieve treatment equivalent to tertiary treatment by more conventional, mechanical means, (2) to establish best management procedures for spray irrigation of such forests in Michigan if irrigation proved feasible, and (3) to measure changes in the vegetation that might result from spray irrigation. Results from the first three years of operation emphasizing a mass balance approach for nitrogen, phosphorus, and chloride have been reported elsewhere (Burton, 1979; Burton and Hook, 1978a, 1978b, 1979; Burton and King, 1979; and King and Burton, 1979). This paper will summarize these earlier studies and will report findings from 1979 and 1980, the last two years of spray irrigation.

DESCRIPTION OF THE STUDY AREA

Three 1.2 ha plots were established in a late successional sugar maple-beech forest in the fall of 1975. These plots were dominated by sugar maple *(Acer saccharum)* (53% dominance of ≥10 cm trunk diameter trees) with beech *(Fagus grandifolia* Ehrh.*)* being less dominant but an important component of the forest (16% dominance of ≥10 cm trunk diameter trees). *Ulmus rubra* (8%), *Tilia americana* (6%), *Ulmus* *americana* (4%), and *Prunus serotina* (4%) were also significant components of the forest (W. Baker and S. Weber, unpublished data available from the author). The forest contained an average of 423 trees/ha $\geq$10 cm diameter at breast height (dbh) with a mean basal area of 42 m^2/ha (Knobloch and Bird, 1978). Detailed analyses of the vegetation are available from Frye (1976a, 1976b, 1977).

The three 1.2 ha plots included a well drained control plot (no irrigation), a well drained area irrigated at the rate of 5 cm/wk during the growing season, and a poorly drained area irrigated at the rate of 10 cm/wk. Soils underlying the two well drained sites were predominantly Miami-Marlette and Kalamazoo loams; well drained members of the mixed mesic family of Typic Hapludalfs formed on glacial till. The Kalamazoo loam contains more sands and gravel than the Miami-Marlette. Soils on the 10 cm/wk site were primarily Owosso sandy loam with some Brookston and Conover loams which were poorly drained. The Owosso sandy loam was well to moderately well drained with a layer of loam to clay loam soil at 45 to 105 cm depths. The site also contained depressions with almost impermeable clays near the surface. Thus, water on this 10 cm/wk site tended to pool in depressions during the spring and early summer, even prior to irrigation. Excessive irrigation was deliberate to try to encourage denitrification as will be discussed in the results. For more detailed descriptions of the site, refer to Burton and Hook (1978a, 1979).

MATERIALS AND METHODS

Analytical methods for the mass balance studies have been described in detail in Burton and Hook (1978a, 1979) and will not be repeated here. Briefly, porous-cup vacuum soil water samplers were installed at depths of 15, 30, 60, 90, 120, and 150 cm at 10 sampling sites for each plot. These soil water samplers were sampled once per week during the growing season during the first three years of the study and once every two weeks during the final two years. Runoff from the 10 cm/wk plot was monitored with an automatic sequential water sampler. Wastewater inputs were calculated from pump records and were sampled for water quality analysis using funnels placed one meter above the ground. Rain inputs were monitored at a nearby location, and evapotranspiration was calculated using the technique of Thornthwaite and Mather (1967).

Wastewater irrigation was accomplished using a fixed set agricultural surface system of aluminum pipes and short risers with agricultural spray nozzles. Wastewater was applied below the canopy at rates of 5 or 10 cm/wk by pumping secondary, chlorinated wastewater from East Lansing, Michigan,

Sewage Treatment Plant, either directly or from the first lake of the WQMF. Application was at the rate of 8.4 mm/hr, two days per week, for total applications of 5 or 10 cm/wk for the two irrigated plots.

Litter fall was sampled using one square meter baskets placed at random just above the forest floor. There were 16 of these baskets in the control plot, 14 in the 5 cm/wk plot, and 10 in the 10 cm/wk plot. Each basket was 15 cm deep and consisted of wooden frames with 6 mm mesh hardware cloth over the bottom. These baskets were sampled throughout the year at intervals of a few days to two or three months depending on intensity of litter fall. Litter was sorted to species, dried, and weighed.

Tree growth studies were conducted after two years of irrigation by coring all trees $\geq$10 cm dbh on the plots and determining growth rates from tree rings.

Soil samples were also taken prior to and after the first year of irrigation in 1976 at sites near the soil water sampler installations.

Water and soil analytical techniques are detailed in Burton and Hook (1978a, 1979).

RESULTS

Mass Balance Studies

The mass balance studies for the first three years of the study demonstrated that older forests on well drained soils in Michigan removed little, if any, nitrogen from secondary municipal wastewater (Burton and Hook, 1978a, 1979). During the first year of application, soil water from the pre-irrigation period slowly came into equilibrium with wastewater as demonstrated by chloride concentrations at various depths. It took almost the entire growing season for this equilibrium to occur (Burton and Hook, 1979). Thus, the first year's data after wastewater application began was not representative of leachate quality from wastewater irrigation. The second and third year data do reflect wastewater irrigation (Table I). The data from the 5 cm/wk well drained site demonstrated that nitrogen applied as inorganic nitrogen was rapidly leached from the site, primarily as nitrate-nitrogen (Table I). In fact, the mass balance for the two year period indicated that inputs almost exactly equalled outputs in soil water leachates (294 kg/ha inputs, 292 kg/ha outputs). Organic nitrogen for the same two year period leached at the rate of 22 kg/ha or 23 percent of input (Table I). Thus, on-site retention of nitrogen was 20 percent of total

Table I

Mass Balances for the 5 cm/wk Irrigated Forest Area, October 1, 1976, to September 30, 1977, and October 1, 1977, to September 30, 1978 (Modified From Burton, 1979, With 1976-77 Data From Burton and Hook, 1978a, 1979).

	*Total Inputs**	*Evapotranspiration*	*Recharge*
Water			
1976-77 (cm/yr)	218.6	66.0	152.6
% of Input	100.0	30.2	69.8
1977-78 (cm/yr)	221.7	63.2	158.5
% of Input	100.0	28.5	71.5
	*Total Inputs***	*Soil Water Leachates*	*On-Site Retention*
Inorganic N			
1976-77 (kg/ha/yr)	172.7	161.8	10.9
% of Input	100.0	93.7	6.3
1977-78 (kg/ha/yr)	121.5	130.6	-9.2
% of Input	100.0	107.6	-7.6
Organic N			
1976-77 (kg/ha/yr)	56.6	9.7	46.9
% of Input	100.0	17.1	82.9
1977-78 (kg/ha/yr)	40.8	12.5	28.3
% of Input	100.0	30.7	69.3
Total N			
1976-77 (kg/ha/yr)	229.3	171.5	57.8
% of Input	100.0	74.8	25.2
1977-78 (kg/ha/yr)	162.3	143.1	19.1
% of Input	100.0	88.2	11.8
Total P			
1976-77 (kg/ha/yr)	43.9	1.5	42.4
% of Input	100.0	3.4	96.6
1977-78 (kg/ha/yr)	44.3	1.3	43.0
% of Input	100.0	2.9	97.1
Chloride			
1976-77 (kg/ha/yr)	1815.6	1869.9	-54.3
% of Input	100.0	103.0	-3.0
1977-78 (kg/ha/yr)	1708.3	1750.7	-42.4
% of Input	100.0	102.5	-2.5

*Includes precipitation inputs of 61.9 and 59.0 cm/yr for 1976-77 and 1977-78, respectively.

**Includes inputs in precipitation plus wastewater.

nitrogen inputs with all of this retention accounted for by retention of organic nitrogen (Table I).

Soil water concentrations were generally the same as wastewater input concentrations with some lag time involved (see Figures in Burton, 1979; Burton and Hook, 1979).

After the results from 1976, 1977, and 1978 indicated that little nitrogen removal was occurring, the feasibility of wastewater irrigation with low nitrogen wastewater from wastewater lagoon systems was investigated. Previous studies had demonstrated that wastewater lagoons were efficient at stripping nitrogen from wastewater (King and Burton, 1979; Burton and King, 1979). Thus, wastewater from the third or fourth lake (lagoon) of the WQMF was spray irrigated on the well drained site at the rate of 5 cm/wk in 1979 and 1980. The site had provided excellent phosphorus removal during the first three years of the study with on-site retention of 97 percent of applied phosphorus (Table I). Thus, wastewater from a lagoon system with low nitrogen and high phosphorus should receive excellent renovation if sprayed on these older forests.

The lag time necessary for equilibration of soil water with incoming wastewater concentrations means that the 1979 data are useful only in following the transition from high nitrogen to low nitrogen wastewater. Data from the last two years of the study are summarized in Table II. These data are calculated by two methods: (1) by using the water budget technique to estimate leachate amounts and (2) by assuming that chloride inputs must equal outputs. The latter assumption results in slightly higher leachate rates (Table II). During the transition period in 1979, leachate losses were more than twice the input of nitrogen (Table II). Losses only slightly exceeded inputs in 1980 (Table II). Even so, losses in 1980 indicated mineralization and loss of some of the applied organic nitrogen or some of the native nitrogen pool. Additions of organic nitrogen to the site more than offset leaching losses of inorganic nitrogen, so wastewater irrigation with low nitrogen wastewater from a lagoon system could be accomplished without contamination of groundwater with concentrations of nitrate-nitrogen in excess of the 10 mg N/l standard (Table III) and without long term losses of productivity due to excessive leaching of native nitrogen from the site.

Phosphorus renovation remained high with input concentrations for 1980 being 1.21 ± 0.85 mg P/l and output concentrations variable but with weekly averages being less than 0.05 mg P/l for most weeks during the 1979 and 1980 growing seasons.

The feasibility of promoting denitrification by excessive irrigation on a poorly drained site was also investigated during the first three years of the study. This technique

Table II

Inorganic Nitrogen Budgets from October 1, 1978, to September 30, 1979, and From October 1, 1979, to September 30, 1980. Water Budget Calculations Used Thornthwaite and Mather's Technique (1967); Chloride Budget Calculations Assumes Chloride Output Must Equal Input.

	1978-79	*1979-80*
Water Budget Basis		
Total Inputs (kg/ha/yr)*	24.5	40.3
Percent of Input	100	100
Groundwater Leaching (kg/ha/yr)	53.5	46.3
Percent of Input	218	115
On-Site Losses (kg/ha/yr)	-29.0	-6.0
Percent of Input	-118	-15
Chloride Budget Basis		
Total Inputs (kg/ha/yr)*	24.5	40.3
Percent of Input	100	100
Groundwater Leaching (kg/ha/yr)	58.4	47.3
Percent of Input	238	117
On-Site Losses (kg/ha/yr)	-33.9	-7.0
Percent of Input	-138	-17

*Includes input of 5.6 and 6.1 kg N/ha/yr in precipitation in 1978-79 and 1979-80 and inputs in 5 cm/wk of wastewater.

Table III

Inorganic Nitrogen Concentrations (mg N/l) in Applied Wastewater and in Soil Water

Irrigation Period	*Wastewater Input*	*Soil Water Leachate*
4/22 to 5/ 7/80	1.80 ± 1.27	2.41 ± 1.40
5/ 8 to 5/21/80	1.21 ± 0.08	3.60 ± 1.92
5/22 to 6/ 4/80	0.75 ± 0.16	3.35 ± 1.44
6/ 5 to 6/18/80	1.06 ± 0.81	3.31 ± 2.24
6/19 to 7/ 2/80	2.54 ± 1.01	2.40 ± 1.23
7/ 3 to 7/17/80	2.20 ± 0.64	2.10 ± 1.81
7/18 to 7/30/80	2.10 ± 0.57	2.58 ± 1.84
7/31 to 8/13/80	3.64 ± 1.12	2.91 ± 2.01
8/14 to 8/27/80	7.30 ± 8.36*	5.54 ± 3.54
8/28 to 9/10/80	2.52 ± 0.44	3.45 ± 2.92
9/11 to 9/24/80	2.66 ± 0.40	4.24 ± 3.47

*Includes one irrigation with wastewater concentration of 19.82 mg N/1.

Table IV

Mass Balances for the 10 cm/wk Irrigated Forest Area, October 1, 1976, to September 30, 1977, and October 1, 1977, to September 30, 1978. (Modified From Burton, 1979, With 1976-77 Data From Burton and Hook, 1978a, 1979).

	*Total Inputs**	*Evapotranspiration*	*Runoff*	*Recharge*
Water				
1976-77 (cm/yr)	375.4	66.0	216.3	92.9
% of Input	100.0	17.6	57.6	24.7
1977-78 (cm/yr)	380.9	63.2	292.9	24.8
% of Input	100.0	16.6	76.9	6.5

	*Total Inputs***	*Runoff*	*Soil-Water Leachate*	*On-Site Retention*
Inorganic N				
1976-77 (kg/ha/yr)	307.6	71.8	20.3	215.6
% of Input	100.0	23.3	6.6	70.1
1977-78 (kg/ha/yr)	226.2	32.1	5.5	188.6
% of Input	100.0	14.2	2.4	83.4
Organic N				
1976-77 (kg/ha/yr)	98.4	66.2	4.2	28.0
% of Input	100.0	67.3	4.3	28.5

1977-78 (kg/ha/yr)	96.7	40.4	0.5	55.8
% of Input	100.0	41.8	0.5	57.7
Total N				
1976-77 (kg/ha/yr)	406.0	138.0	24.4	243.6
% of Input	100.0	34.0	6.0	60.0
1977-78 (kg/ha/yr)	322.9	72.5	6.0	244.4
% of Input	100.0	22.5	1.9	75.7
Total P				
1976-77 (kg/ha/yr)	86.8	29.0	0.6	57.2
% of Input	100.0	33.4	0.7	65.9
1977-78 (kg/ha/yr)	83.0	37.5	0.04	45.4
% of Input	100.0	45.2	0.05	54.7
Chloride				
1976-77 (kg/ha/yr)	3750.7	2305.4	1080.9	364.4
% of Input	100.0	61.5	28.8	9.7
1977-78 (kg/ha/yr)	3420.3	3006.9	238.8	174.6
% of Input	100.0	87.9	7.0	5.10

*Includes precipitation inputs of 61.9 and 59.0 cm/yr for 1976-77 and 1977-78, respectively.

**Includes inputs in precipitation plus wastewater.

worked well for preventing soil water leachate losses (Table IV) with soil water concentrations of inorganic nitrogen decreasing from a high of about 4 mg N/l during the spring to concentrations below 1 mg N/l late in the summer (Burton, 1979; Burton and Hook, 1979). However, excessive runoff did occur resulting in poor phosphorus renovation (55 to 66% removal) and reductions in total nitrogen renovation to about 68 percent of input during a two year period (Table IV). It might be possible to apply just enough water to keep the soil saturated but not enough to result in excessive runoff so that excellent renovation of wastewater would occur. However, this management procedure would be very site specific and time consuming for the operator. Also, standing water in the depressions resulted in death of almost half the trees in this plot during the third year of operation. Thus, this technique is not recommended.

Vegetation Responses

All the trees on the plots were cored and growth increments were determined for the past twelve years at the end of 1977 after two years of irrigation (10 years pre-spray, 2 years irrigation) for an independent study project (W. Baker and S. Weber, unpublished data available from the author). This study demonstrated no significant differences in growth between the control and either the 5 or 10 cm/wk irrigated areas. This lack of difference between the control and 10 cm/wk site was especially surprising, since almost half the trees on the 10 cm/wk site failed to produce leaves in 1978. Death of these trees occurred after the 1977 growing season. Growth determination from tree cores has an inherent high variability which apparently masked differences in growth. There did appear to be a negative trend in growth for *Acer saccharum* and *Fagus grandifolia* and a positive trend in growth for *Ulmus rubra* for the 5 and 10 cm/wk sites compared to the controls, but these trends were not significantly different ($p \leq 0.05$).

Litter production in the forest was also studied during 1977. Litter fall was 5952 ± 1357 kg/ha/yr for the control site, 5388 ± 762 for the 5 cm/wk site, and 4562 ± 1379 kg/ha/yr for the 10 cm/wk site. Litter fall was not significantly different between the control and 5 cm/wk plot but was significantly different for the 10 cm/wk site compared to the control ($p \leq 0.02$ using the t-test for unequal sample size). Further, litter fall began earlier on the 10 cm/wk site with substantial litter fall starting as early as September 1 and with peak litter fall occurring from October 4 to 18, 1977. Eighty-five percent of litter fall had occurred by October 18 for the 10 cm/wk site but only 61 and 57 percent had occurred on the control and 5 cm/wk sites, respectively. Peak litter fall on the latter two sites occurred from October 15 to November 1, 1977, with 36 to 40 percent of annual litter fall occurring in the 10 day period from October 21 to November 1 after 86 percent of annual litter fall had already occurred on the 10 cm/wk site.

The changes in the herbaceous vegetation were also sampled in detail in 1976 and 1977, but calculations on these data are incomplete. Qualitatively, no differences were observed between the control and the 5 cm/wk site. The 10 cm/wk site remained relatively unchanged through 1977. However, an explosive growth of herbaceous vegetation did occur after death of the trees from 1978-1980. There were large monotypic stands of *Impatiens capensis* and substantial increases in thistle (*Cirsium* sp.) and several other species of herbs in the openings created by death of the trees.

Changes in vegetation resulting from wastewater irrigation can be summarized as follows. There were no substantial differences in growth, litter fall, or herbaceous vegetation observed between the control and the 5 cm/wk irrigated plot. The 10 cm/wk irrigated plot produced significantly less litter in 1977 than did the control or the 5 cm/wk plots and about half the trees on this site died after the 1977 growing season before the start of the third season of irrigation. Litter fall occurred earlier on the 10 cm/wk plot in 1977 than on the other plots. Herbaceous growth on the 10 cm/wk plot increased substantially following death of many of the trees.

CONCLUSIONS AND RECOMMENDATIONS

Irrigation with secondary municipal effluent in older forests in Michigan will result in excessive leaching of nitrate to groundwater. However, irrigation with low nitrogen wastewater from a lagoon results in excellent phosphorus removal. Thus, older forests should be used for wastewater irrigation only if low nitrogen wastewater is available. They do offer significant potential for phosphorus removal from wastewater lagoon systems which do not currently meet phosphorus standards for discharge. If irrigation rate is limited to the hydraulic capacity of the soil or to 5 cm/wk or less, no significant impacts on vegetation growth or community structure should occur based on the studies reported here.

ACKNOWLEDGMENT

This work was supported by Grant R005143-01 from the U.S. Environmental Protection Agency and by funds from

Michigan State University. I thank Charles Annett, Paul Bent, William Baker, Joe Ervin, Bobby Holder, William Larsen, Dan O'Neill, John Przybyla, and Steve Weber for field and technical assistance. Many students too numerous to name also participated in these studies. Special thanks are due James E. Hook who was co-investigator for the first three years of this project.

LITERATURE CITED

Burton, T.M. 1979. Land Application Studies on the Water Quality Management Facility at Michigan State University. In Proc. Second Annual Conference Applied Research Practice on Municipal and Industrial Waste, September 17-21, Madison, WI, pp. 112-128.

Burton, T.M. and J.E. Hook. 1979. A Mass Balance Study of Application of Municipal Waste Water to Forests in Michigan. J. Environ. Qual. 8:589-596.

Burton, T.M. and J.E. Hook. 1978a. Application of Municipal Wastewater to Forest Lands. In T.M. Burton, The Felton-Herron Creek, Mill Creek Pilot Watershed Study. EPA-905/9-78-002, U.S. Environmental Protection Agency, Region V, Chicago, IL 60604, pp. 86-120.

Burton, T.M. and J.E. Hook. 1978b. Use of Natural Terrestrial Vegetation for Renovation of Wastewater in Michigan. In H.L. McKim (ed.), Proc. State of Knowledge in Land Treatment of Wastewater, U.S. Army Cold Regions Research Laboratory, Hanover, NH 03755, August 20-25, 2:199-206.

Burton, T.M. and D.L. King. 1979. A Lake-Land System for Recycling Municipal Wastewater. In Proc. 1979 National Conference on Environmental Engineering, July 9-11, ASCE, San Francisco, CA, pp. 68-75.

Frye, D.M. 1976a. A Botanical Inventory of Sandhill Woodlot, Ingham County, Michigan. I. The Vegetation. Mich. Bot. 15:131-140.

Frye, D.M. 1976b. A Botanical Inventory of Sandhill Woodlot, Ingham County, Michigan. II. Checklist of Vascular Plants. Mich. Bot. 15:195-204.

Frye, D.M. 1977. A Botanical Inventory of Sandhill Woodlot, Ingham County. III. Phenology Study. Mich. Bot. 16:34-38.

King, D.L. and T.M. Burton. 1979. A Combination of Aquatic and Terrestrial Ecosystems for Maximal Reuse of Domestic Wastewater. In Proc. Water Reuse - From Research to Application, March 25-30, 1979, Washington, D.C., Amer. Water Works Assoc. Res. Found., Denver, CO 80235, 1:714-726.

Knobloch, N. and G.W. Bird. 1978. Criconematinae Habitats and *Lobocriconema thornei* n. sp. (Criconematidae: Nematoda). J. Nematol. 10:61-70.

Thornthwaite, D.W. and J.R. Mather. 1967. Instructions and Tables for Computing Potential Evapotranspiration and the Water Balance. Climatology 10:185-311. (Laboratory of Climatology, Drexel Institute of Technology, Centerton, NJ.)

CHAPTER 12

PLANT DISEASES ASSOCIATED WITH MUNICIPAL WASTEWATER IRRIGATION*

Lynn Epstein and Gene R. Safir
Department of Botany and Plant Pathology
Michigan State University
East Lansing, Michigan 48824

INTRODUCTION

Disposal of secondary treated wastewater onto land is frequently a desirable alternative to direct disposal into surface waters (Bower and Chaney, 1974; D'Itri, 1977). The land acts as a living filter and can often provide an economical means of tertiary cleanup. However, the possibility that human pathogens could be introduced with the wastewater raises additional concerns.

Four factors determine the interaction of wastewater irrigation and plant disease in an old field. First, since germination and growth of many plant pathogens is favored by both high soil moisture and nutrient levels, the long term efficiency or stability of the wastewater irrigated ecosystem could possibly be threatened by disease. Secondly, the buildup of plant disease in an old field could pose a threat to neighboring fields of economically important alternate hosts. Thirdly, it is possible that municipal wastewater may contain plant pathogens (Cooke, 1956). Fourthly, the wastewater could affect the survival of several organisms.

PLANT DISEASE INCIDENCE

Most reports on damage to vegetation from applied wastewater have emphasized the role of salts or overfertilization (Baier and Fryer, 1973), particularly when the water

*Michigan Agricultural Experiment Station Journal Article No. 9793.

is derived from agricultural and food industry wastes (Morisot and Gras, 1974). Neary *et al.* (1975) found evidence of boron toxicity to red pine needles in a plantation irrigated with 2.5, 5 and 8.8 cm of wastewater/wk. Other authors have reported tree death due to ice damage (Sopper and Kardos, 1972).

Relatively few papers in the field of wastewater irrigation discuss plant disease. In one study, an unidentified root rot complex (Marten *et al.*, 1979) was involved in a severe decline of alfalfa in wastewater irrigated plots at 5 and 10 cm/wk applications. The root rot first developed by the second year after seeding and by the fifth year, percent stand in the 0, 5, and 10 cm/wk irrigated plots was 84, 28, and 2 percent, respectively. In this study the plants were harvested three times per year and the controls received well water and a mineral fertilizer when moisture stress appeared. Zeiders (1975) and Zeiders and Sherwood (1977) found significantly more tawny blotch (caused by *Stagonospora foliicola*) on reed canarygrass irrigated with 5 cm municipal wastewater than on non-irrigated plants. A three-cutting harvest system significantly reduced disease, and clones of reed canarygrass varied from resistant to susceptible.

The percent chlorotic or necrotic tissue of the leaves visible in the canopy were assessed in a Michigan State University wastewater irrigated old field. Data were taken on 1 date in 1977, 4 dates in 1978, and 5 dates in 1979. Three irrigation treatments (0, 5, and 10 cm wastewater/wk) and three harvest regimes (0, 1, or 2 harvests/season) were utilized. There were four replicate plots of each irrigation-cutting combination. The dominant plants were goldenrod (*Solidago* spp.) in the unharvested plots and quackgrass (*Agropyron repens*) in the harvested plots. By the end of the season in 1977, 1978, and 1979, there was significantly more disease in the uncut irrigated plots than in the cut, irrigated or any of the non-irrigated plots. However, the marked differences in uncut versus cut plots appeared relatively late in the season. By September 1 in 1978 and 1979, less than 5 percent of the leaf tissue in harvested plots was diseased in either non-irrigated or irrigated plots. In 1978, approximately 10 percent of the leaf tissue in the uncut, 5 cm wastewater irrigated and 20 percent of the leaf tissue in the uncut, 10 cm wastewater irrigated plots were diseased. In 1979, approximately 15 percent of the leaf tissue was diseased in both the 5 cm and 10 cm wastewater uncut treatment. On June 30 and August 24, 1978, we sampled root tissue to estimate the percent of necrotic tissue and found no significant differences among the treatments.

The diseases identified in the old field were predominately of fungal origin. *Coleosporium asterum* and *Erysiphe cichoracearum* caused a rust and a powdery mildew on goldenrod;

Helminthosporium sp. and *Phyllachora graminis* caused leaf spots on quackgrass. No significant bacterial diseases were detected. There was one virus-like disease on goldenrod, characterized by ringspot and oak leaf patterns on the leaves. The disease was observed in approximately half of the irrigated plots, but not in any of the unirrigated plots.

In cooperation with Michigan State University nematologist Dr. George Bird, the soil was sampled for the presence of plant parasitic nematodes on September 7, 1977, and July 6, 1978. In both cases, fewer plant parasitic nematodes were found in the irrigated than in the non-irrigated plots (20, 42, and 155 plant parasitic nematodes/100 cc soil in 1977 and 7, 29, and 98 plant parasitic nematodes/100 cc soil in 1978 for the 10, 5, and 0 cm/wk irrigation treatments, respectively).

There has been several studies of the effects of sludge, particularly manure, on soilborne plant diseases. Sludge compost added to soils (10% w/w) decreased *Aphanomyces* root rot of peas (75-80%), *Rhizoctonia* damping-off of cotton (up to 50%), and *Sclerotinia* drop of lettuce (up to 50%). However, sludge compost did not affect four other soilborne diseases; occasionally diseases caused by *Pythium* and *Fusarium* in pea and *Thielavioposis* in beans were increased (Lumsden *et al*., 1980). Dazzo (1972) reported a reduction in damping off of oat, sorghum, and millet caused by *Pythium aphanidermatum*, in plots treated with slurried dairy manure. McIlveen and Cole (1977) found an increased incidence and severity of *Gibberella* ear rot of corn in plots receiving increased sludge or manure treatments; Stewart's bacterial wilt was unaffected.

AGRONOMIC PLANT DISEASES

Non-agricultural plants are an integral part of the life cycle of many phytopathogenic organisms (Dinoor, 1974). Numerous economically important plant pathogens can propagate on alternate weed hosts. In addition to being sources of inoculum, perennial weeds could serve as overwintering hosts for plant pathogens. According to the Index of Plant Diseases (Anon., 1960) and the Handbook of Plant Disease (Westcott, 1971), about twenty pathogens of goldenrod and almost thirty pathogens of quackgrass can also cause disease on economically important hosts. Quackgrass can be infected with such diseases as ergot in rye, take all in wheat, and *Septoria* leaf blotch in numerous cereals and grasses (Sprague, 1960). There were two major goldenrod diseases responsible for the disease increase in the old field in the uncut irrigated plots. The rust disease on goldenrod is caused by *Coleosporium asterum*, a fungus which alternates between goldenrod or aster and numerous pine species (Nicholls *et al*., 1968). The powdery mildew disease on goldenrod is caused by the pathogen *Erysiphe cichoracearum*. Although the genetics of *E. cichoracearum* has not been studied closely, a host range of nearly 300 plant hosts has been reported (Westcott, 1971).

A limited number of studies have reported minimal detrimental effects of wastewater on populations of soil bacteria (Ellis *et al.*, 1979; Cairns *et al.*, 1978), actinomycetes (Orchard, 1978), and protozoans (Stout, 1978). However, earthworms were suppressed by consistently wet conditions (Stout, 1978). In a study on the effect of sludge on microbial populations, McIlveen and Cole (1977) found increased populations of bacteria and actinomycetes in soil treated with 22 and 44 metric tons of sludge/ha. There were only small changes in fungal populations. Dazzo *et al.* (1974) reported microbiota in soil irrigated with slurried dairy manure were similar to normal rhizosphere populations.

PLANT PATHOGENIC PROPAGULES

Cooke (1956) reported the presence of "potentially" plant pathogenic fungi in municipal sewage. The "fungi of interest to the plant pathologist" include *Alternaria tenuis*, 4 *Aspergillus* spp., *Botrytis cinerea*, *Cephalosporium* spp., 2 *Chaetomium* spp., *Cladosporium cladosporioides*, *Coniothyrium fuckelii*, *Curvularia lunata*, 5 *Fusarium* spp., 2 *Gliocladium* spp., *Mucor* spp., 9 *Penicillium* spp., *Rhizopus nigricans*, *Scopulariopsis brevicaulis*, *Stemphyllium consortiale*, and *Trichoderma viride*. Some of the fungi in the class Oomycetes (most notably *Phytophthora* sp.) occur in recycled irrigation water (Thomson and Allen, 1976) and are quite well suited for survival in an aqueous environment. For example, microsporangia of *Phytophthora parasitica* survived for 60 days at 24°C in irrigation wastewater (Thomson and Allen, 1976), and McIntosh (1966) isolated *Phytophthora* sp. in 27 out of 31 irrigation sources in Canada.

In 1978 and 1979, injured seedlings were used as baits for plant pathogens in the East Lansing wastewater. This method has two advantages over pure culture methods for detecting pathogenic organisms: (1) only pathogenic microorganisms are recovered, and (2) water samples need not be concentrated. The plants used as baits included two cultivars each of corn, barley, soybeans, oats, wheat and alfalfa, one cultivar of navy bean, and two collections of quackgrass. No pathogens were recovered from the wastewater in this study.

There are several possible explanations why plant pathogenic organisms were not isolated from the East Lansing wastewater. First, the majority of "plant pathogenic" species isolated by Cooke (1956) are only weak parasites. *Alternaria tenuis (A. alternata)*, for example, is a general saprophyte, and only occasionally a weak parasite (Horst, 1979). Other weak parasites isolated by Cooke, associated with debilitated or wounded tissue, include *Aspergillus* sp., *Mucor* sp., *Penicillium* sp., and *Rhizopus* sp. While *Fusarium* spp. cause many important rots, wilts and yellows diseases, non-pathogenic isolates are extremely common. Second, phytopathogenic organisms may be present only in very small populations. While fungi can be routinely isolated from wastewater treatment plants (Cooke, 1976), flocs do not usually contain large amounts of fungal material (LaRiviere, 1977). Of the 10 bacterial genera commonly reported as floc constituents (Pike, 1975) only one genus *(Pseudomonas)* contains any phytopathogenic species. Any phytopathogenic viruses present in the wastewater are probably of little consequence. In order for a plant virus infection to occur, the virion must apparently contact a wounded cell. In addition, most plant viruses are transmitted by a specific vector. Third, chlorination of wastewater may have killed phytopathogenic organisms. Unfortunately, little is known concerning the question as to whether chlorination kills plant pathogenic propagules. However, routine waterings of mushroom beds with chlorinated water is an effective means of controlling brown blotch on mushrooms, caused by the bacterium *Pseudomonas tolaasi* (Royse and Wuest, 1980). Fungi may be considerably more resistant to chlorine than bacteria. The yeast *Candida parapsilosis*, a human pathogen, is more resistant to chlorine than human bacillary pathogens and waterborne viruses (Engelbrecht *et al.*, cited in White, 1978). While it seems unlikely there are high levels of phytopathogenic organisms in the East Lansing wastewater, there may be a greater number of pathogenic propagules in effluent with substantial vegetative wastes.

GROWTH AND SURVIVAL OF PLANT PATHOGENS

Many fungi can be stored in water for relatively long periods of time. We tested the survival of two fungi, *Alternaria alternata* and *Stemphylium sarcinaeforme* in wastewater and tap water. *Alternaria alternata* is one of Cooke's (1956) "potentially plant pathogenic" sewage fungi. Conidia of the two fungi were inoculated into filter sterilized wastewater or tap water. Periodically over a 24 day period, 0.1 ml aliquots of wastewater or tap water were plated onto potato dextrose agar in order to monitor populations. There were no differences in the survival of the spores in tap and wastewater. Neither fungus produced detectable vegetative growth in either water. The wastewater used may have been too dilute, especially with respect to sources of carbon, to support extensive microbial growth.

It is believed plant parasites and saprophytes compete for nutrients in exudates found on plant surfaces. To examine nteraction, we monitored the short term survival of a athogenic bacterium, *Erwinia herbicola*, in wastewater and tap water. A sterile dilute phosphate buffer was added to filter-sterilized wastewater or tap water with or without 0.1 percent yeast extract (to simulate plant exudate). After inoculation with a uniform bacterial suspension containing approximately 10^5 cells/ml, the flasks were shaken for 24 hr. Populations at 0 and 24 hr after inoculation were determined with a dilution series of the liquid on yeast extract agar. Both the pathogenic and the saprophytic *Erwinia* species survived better in the wastewater than in the tap water, but grew more rapidly in the simulated leaf exudate in tap water than in simulated leaf exudate in wastewater.

CONCLUSIONS

Plant disease only occurs when a pathogen is present, when the host is of susceptible genotype and the environment favors disease development. In wastewater irrigated systems, the greatest disease problems may be associated with *Pythium* and *Phytophthora* root rots. The pathogens are widespread and in general appear to be most prevalent in wet soils. *Phytophthora* root rot of alfalfa can be a serious problem in poorly drained areas. Soil saturation prior to inoculation predisposes alfalfa to *Phytophthora* root rot (Kuan and Erwin, 1980) and subsequent additional water enhances disease development (Pulli and Tesar, 1975). Alfalfa has been recommended (Almy *et al.*, 1977) as a suitable crop for land irrigated with wastewater. On the basis of the findings of Marten *et al.* (1979) and our present information of the disease (Frosheiser, 1969), a degree of caution should be exercised when planting alfalfa in imperfectly drained areas irrigated with wastewater.

In conclusion, significant differences between the wastewater and tap water that would be of plant pathological significance were not observed. Assuming the wastewater neither contains pathogens nor serves as a nutrient source for microbes, then the primary effects of wastewater irrigation on plant disease are most likely the same as those caused by irrigation alone. Wastewater irrigation project managers should be aware that excessive moisture generally promotes plant disease, that an epiphytotic could seriously decrease the ability of the vegetation to remove nutrients, and that an uncultivated area with high plant disease can serve as a source of inoculum and as a reservoir for pathogens which can attack economically important hosts.

LITERATURE CITED

Almy, A.A., P.A. Blakeslee, L.J. Connor, D.R. Isleib, L.W. Jacobs, L.W. Libby, T.L. Loudon, R.E. Nelson, L.D. Stephens, and E.T. Van Nierop. 1977. Land Application of Municipal Wastewater. Michigan State University Extension Bulletin E-1138 Natural Resources Series. Cooperative Extension Service, Michigan State University, East Lansing, MI 48824, 4 pp.

Anon. 1960. Plant Pests of Importance to North America Agriculture. U.S. Printing Office, Washington, D.C., 531 pp.

Baier, D.C. and W.B. Fryer. 1973. Undesirable Plant Response With Sewage Irrigation. Amer. Soc. Civil Eng. Irrigat. Drain. Div. J. 99:133-141.

Bower, H. and R.L. Chaney. 1974. Land Treatment of Wastewater. Adv. Agron. 26:133-176.

Cairns, A., M.E. Dutch, E.M. Guy, and J.D. Stout. 1978. Effect of Irrigation With Municipal Water or Sewage Effluent on the Biology of Soil Cores. I. Introduction. Microbial Populations and Respiratory Activity. N. Z. J. Agric. Res. 21:1-9.

Cooke, W.B. 1956. Potential Plant Pathogenic Fungi on Sewage and Polluted Water. Plant Dis. Reptr. 10:681-687.

Cooke, W.B. 1976. Fungi in Sewage. In E.G. Jones (ed.), Recent Advances in Mycology. Halsted Press, NY, pp. 289-434.

Dazzo, F.B. 1972. The Microbial Ecology of Cultivated Soil Receiving Cow Manure Waste. M.S. Thesis, University of Florida, Gainesville, FL, 97 pp.

Dazzo, F.B., P.H. Smith, and D.H. Hubbell. 1974. Changes in the Rhizosphere Effect of Millet Associated With Sprinkler Irrigation With Animal Wastes. J. Environ. Qual. 3: 270-273.

Dinoor, A. 1974. Role of Wild and Cultivated Plant Diseases in Israel. Ann. Rev. Phytopathol. 12:413-436.

D'Itri, F.M. (ed.) 1977. Wastewater Renovation and Reuse. Marcel Dekker, Inc., New York, NY, 705 pp.

Ellis, B.G., A.E. Erickson, A.R. Wolcott, B.D. Knezek, J.M. Tiedje, and J. Butcher. 1979. Applicability of Land Treatment of Wastewater in the Great Lakes Area Basin: Effectiveness of Sandy Soils at Muskegon Co., Michigan, for Renovating Wastewater. EPA 905/79-006-B, U.S. Environmental Protection Agency, Region V, Chicago, IL 60604.

Frosheiser, F.I. 1969. *Phytophthora* Root Rot of Alfalfa in the Upper Midwest. Plant Dis. Rep. 53:595-597.

Horst, R.K. 1979. Wescott's Plant Disease Handbook. 4th Ed. Van Nostrand, New York, NY, 803 pp.

Kuan, T.L. and D.C. Erwin. 1980. Predisposition Effect of Water Saturation of Soil on Phytophthora Root Rot of Alfalfa. Phytopathology 70:981-986.

LaRiviere, J.W.M. 1977. Microbial Ecology of Liquid Waste Treatment. In M. Alexander (ed.), Advances in Microbial Ecology. Plenum Press, New York, NY, 1:215-259.

Lumsden, R.D., J.A. Lewis, R.E. Werner, and P. Millner. 1980. Effect of Sludge Compost on Selected Soilborne Diseases. Phytopathology 71:238 (Abstr.).

McIlveen, W.D. and H. Cole, Jr. 1977. Influence of Sewage Sludge Soil Amendment on Various Biological Components of the Corn Field Ecosystem. Agriculture and Environment 3:349-361.

McIntosh, D.L. 1966. The Occurrence of *Phytophthora* spp. in Irrigated Systems in British Columbia. Can. J. Bot. 44:1591-1596.

Marten, G.C., C.E. Clapp, and W.E. Larson. 1979. Effects of Municipal Wastewater Effluent and Cutting Management on Persistence and Yield of Eight Perennial Forages. Agron. J. 71:650-658.

Morisot, A. and R. Gras. 1974. Agronomic Effects of the Land Disposal of Wastes from the Agricultural and Food Industries. Annales Agronomiques 25:243-266.

Neary, D.G., G. Schnider, and D.P. White. 1975. Boron Toxicity in Red Pine Following Municipal Wastewater Irrigation. Soil Sci. Soc. Amer. Proc. 39:981-982.

Nicholls, T.H., R.F. Patton, and E.P. VanArsdel. 1968. Life Cycle and Seasonal Development of *Coleosporium* Pine Needle Rust in Wisconsin. Phytopathology 58:822-829.

Orchard, V.A. 1978. Effect of Irrigation With Municipal Water or Sewage Effluent on the Biology of Soil Cores. III. Actinomycete flora. N. Z. J. Agric. Res. 21:21-28.

Pike, E.B. 1975. Aerobic Bacteria. In C.R. Curds and H.A. Hawkes (eds.), Ecological Aspects of Used-Water Treatment. Academic Press, London, 1:1-63.

Pulli, S.K. and M.B. Tesar. 1975. *Phytophthora* Root Rot in Seedling-Year Alfalfa as Affected by Management Practices Inducing Stress. Crop Sci. 15:851-854.

Royse, D.J. and P.J. Wuest. 1980. Mushroom Brown Blotch: Effects of Chlorinated Water on Disease Intensity and Bacterial Populations in Casing Soil on Pilei. Phytopathology 70:902-905.

Sopper, W.E. and L.T. Kardos. 1972. Vegetation Responses in Irrigation With Treated Municipal Wastewater. In W.E. Sopper and L.T. Kardos (eds.), Recycling Treated Municipal Wastewater and Sludge Through Forest and Cropland. The Pennsylvania State University Press, University Park, PA 16802, pp. 271-294.

Sprague, R. 1960. Diseases of Cereals and Grasses in North America. The Ronald Press, NY, 538 pp.
Stout, J.D. 1978. Effect of Irrigation With Municipal Water or Sewage Effluent on the Biology of Soil Cores. II. Protozoan faunz. N. Z. J. Agric. Res. 21:11-20.
Thomson, S.V. and R.M. Allen. 1976. Mechanisms of Survival of Zoospores of *Phytophthora parasitica* in Irrigation Water. Phytopathology 66:1198-1202.
Westcott, C. 1971. Plant Disease Handbook. 3rd Ed. D. Van Nostrand Co., New York, NY, 782 pp.
White, G.C. 1978. Disinfection of Wastewater and Water for Reuse. Van Nostrand Reinhold Co., New York, NY, 387 pp.
Zeiders, K.E. 1975. *Stagonospora foliicola*, A Pathogen of Reed Canary Grass Spray-Irrigated With Municipal Sewage Effluent. Plant Dis. Rep. 59:779-783.
Zeiders, K.E. and R.T. Sherwood. 1977. Effect of Irrigation With Sewage Wastewater, Cutting Management, and Genotype on Tawny Blotch of Reed Canary Grass. Crop Sci. 17:594-596.

AUTHOR INDEX

A

Aldrich, S.R., 49
Allen, R.M., 198
Almy, A.A., 200
Aly, O.M., 30
Anon., 197

B

Baier, D.C., 195
Baker, W., 182, 190
Barnes, R.F., 147
Barr, H.L., 147
Bar-Yosef, B., 49
Bates, D.J., 30
Bengtson, G.W., 170
Benne, E.J., 104
Bird, G.W., 182, 197
Blakeslee, P.A., 200
Blank, G.N., 105
Bourget, S.J., 104
Bower, H., 195
Brendemuehl, R.H., 170
Brister, G.H., 155, 158
Brockway, D.G., 5, 165, 166, 172, 174
Brown, B.A., 104
Burton, T.M., 5, 76, 107, 108, 109, 111, 112, 118, 119, 120, 121, 122, 123, 124, 125, 126, 127, 129, 177, 181, 182, 183, 184, 185, 188, 190
Butcher, J., 198

C

Cairns, A., 198
Chaney, R.L., 195
Clapp, C.E., 46, 47, 69, 71, 72, 73, 74, 75, 104, 140, 196, 200
Cole, D.W., 157
Cole, H., Jr., 197, 198
Connor, L.J., 200
Cooke, W.B., 195, 198, 199
Cooley, J.H., 5, 155, 157, 159, 161, 168, 170

D

Dazzo, F.B., 197, 198
Deemer, D.D., 30
Dexter, S.T., 143

Dils, R.E., 155
Dinoor, A., 197
D'Itri, F.M., 5, 195
Dow, B.K., 104
Dowdy, R.H., 5, 46, 47, 140
Dutch, M.E., 198

E

Einspahr, D.W., 170
Ellis, B.G., 5, 49, 54, 55, 56, 57, 58, 59, 60, 61, 62, 63, 67, 70, 71, 198
Epstein, L., 5, 110, 195
Erickson, A.E., 5, 49, 54, 55, 56, 57, 58, 59, 60, 61, 62, 63, 67, 70, 71, 198
Erwin, D.C., 200
Evans, J.O., 165, 170

F

Faisst, J.A., 30, 32
Finn, B.M., 104
Fisk, D.J., 30, 136
Frosheiser, F.I., 200
Frye, D.M., 182
Fryer, W.B., 195

G

Gardner, W.R., 104
Gaskin, D.A., 30, 136
Gilde, L.C., 30
Gilley, J.R., 47
Gorrill, A.D.L., 105
Graham, J.M., 147, 148
Gras, R., 196
Gupta, S.C., 46
Guy, E.M., 198

H

Hall, D.H., 31
Hansen, W.F., 170
Hardee, M.L., 170
Harris, A.R., 159
Hinrichs, D.J., 30, 32
Hook, J.E., 5, 49, 54, 55, 56, 57, 58, 59, 60, 61, 62, 63, 65, 66, 67, 70, 71, 73, 75, 76, 79, 107, 108, 109, 111, 112, 114, 117, 181, 182, 183, 184, 185, 188, 190
Horst, R.K., 198
Howe, J.P., 170

I

Ingalls, J.R., 104
International Joint Commission, 116
Isleib, D.R., 200

J

Jacobs, L.W., 5, 49, 54, 55

56, 57, 58, 59, 60, 61, 62, 63, 67, 70, 71, 200
Jenkins, T.F., 30, 32, 135, 136, 137, 138, 140, 143, 146

K

Kadlec, R.H., 29
Kafkafi, U., 49
Kardos, L.T., 24, 66, 72, 73, 104, 105, 155, 157, 158, 170, 196
Kaufman, M.R., 170
Kerr, S.N., 5, 157
King, D.L., 107, 117, 157, 181, 185
King, E.D., 31
Knezek, B.D., 5, 49, 54, 55, 56, 57, 58, 59, 60, 61, 62, 63, 67, 70, 71, 79, 198
Knobloch, N., 182
Kozlowski, T.T., 171
Kramer, P.J., 171
Kresge, C.B., 143
Kuan, T.L., 200

L

LaRiviere, J.W.M., 199
Larson, R.E., 47
Larson, W.E., 46, 47, 69, 71, 72, 73, 74, 75, 104, 140, 196, 200
Law, J.P., 30
Lawrence, C.H., 31
Lee, C.R., 29, 30, 136
Leggett, D.C., 143
Leland, D.E., 107, 108, 109, 177
Lewis, J.A., 197
Libby, L.W., 200
Liegel, E.A., 138
Linden, D.R., 46, 47, 69, 71, 72, 75, 104
Little, S., 155
Loudon, T.L., 200
Lull, H.W., 155
Lumsden, R.D., 197

M

Malhotra, S.K., 27
Martel, C.J., 30, 32, 135, 136, 140, 143
Marten, G.C., 46, 47, 71, 72, 73, 74, 75, 104, 140, 196, 200
Mather, J.R., 109, 182, 186
McCormack, D.E., 20
McIlveen, W.D., 197, 198
McIntosh, D.L., 198
McKenzie, R.E., 104
McKim, H.L., 30, 136, 140
McPherson, J.B., 136
Melbourne Board of Works Farm, 104
Mill, R.A., 31

Miller, R.H., 104
Millner, P., 197
Morisot, A., 196
Munsell, R.L., 104
Myers, E.A., 5, 19, 25, 27, 28, 31, 72
Myers, L.H., 30

N

Neary, D.G., 157, 159, 170, 171, 177, 196
Neller, J.R., 104
Nelson, R.E., 200
Nesbitt, J.B., 72
New England Crop and Livestock Reporting Board, 145
Nicholls, T.H., 197
Nielsen, K.F., 104
Nutter, W.L., 155, 158
Nylund, J.R., 46, 47

O

O'Donovan, P.B., 104
Orchard, V.A., 198
Otto, P.C., 158

P

Pair, C.H., 24
Palazzo, A.J., 5, 30, 32, 46, 135, 136, 137, 138, 140, 143, 146, 147, 148
Parizek, R.R., 24, 72
Patton, R.F., 197
Peters, R.E., 29, 30, 136
Pike, E.B., 199
Pivetti, D.A., 30, 32
Pollock, T.E., 29, 30
Poloncsik, S., 5
Post, D.M., 170
Powell, R.D., 105
Pritchett, W.L., 170
Pulli, S.K., 200

R

Reed, F., 118, 119, 125
Remson, I., 155
Royse, D.J., 199
Rudolf, P.O., 155
Rudolph, V.J., 155, 159

S

Safir, G.R., 5, 110, 195
Schiess, P., 157
Schneider, G., 157, 159, 166, 171, 172, 174, 196
Schroeder, E.D., 30, 32
Schulte, E.E., 138
Schultz, R.C., 155, 158
Settergren, C.D., 170
Shaffer, M.J., 46
Shelton, J.E., 31
Sherwood, R.T., 196

Smith, W.H., 165, 170
Sopper, W.E., 5, 24, 72, 104, 155, 157, 158, 170, 196
Sprague, R., 197
Staubus, J.R., 147
Stephens, L.D., 200
Stout, J.D., 198
Sutherland, J.C., 5, 29, 170

T

Tesar, M.B., 5, 66, 67, 79, 104, 105, 132, 200
Thomas, J.W., 104, 105
Thomas, R.E., 1, 5, 30
Thomson, S.V., 198
Thornthwaite, C.W., 136
Thornthwaite, D.W., 109, 182, 186
Tiedje, J.M., 198
Trouse, A.C., Jr., 22
Turner, J.A., 170

U

U.S. Department of Agriculture, Agricultural Research, 47
U.S. Environmental Protection Agency, 29, 30, 31, 53
Urie, D.H., 5, 170

V

VanArsdel, E.P., 197
VanNierop, E.T., 200

W

Weber, S., 182, 190
Werner, R.E., 197
Westcott, C., 197, 198
White, D.P., 157, 159, 166, 171, 172, 174, 196
White, G.C., 199
Wiggert, D.C., 107, 108, 109, 177
Williams, T.C., 29
Wolcott, A.R., 198
Wuest, P.J., 199

Z

Zeiders, K.E., 196

INDEX

Abies balsamea, 156, 160, 163
Acer rubrum, 158-159, 191
Acer saccharum, 158-160, 190
Actinomycetes, 198
Aerosol drift, 6
Agropyron repens L., 13, 37, 41, 52, 62, 71, 73, 88, 89, 107, 108, 118, 121, 128, 136, 140, 143, 151, 166, 196-198
Alfalfa, 10, 12-13, 37, 41, 52, 61, 71-75, 81, 83-86, 88-92, 94, 97-99, 101-103, 196, 198, 200
Alopecurus arundinaceus Poir, 81
Alternoria tenuis, 198
Alternaria alternata, 198
Aluminum, 83, 94
Ammonia, 29, 32, 37, 38, 42, 44, 115, 138, 161, 167
Andropogon, 166
Animal
  damage, 67, 83, 161
  feed, 102, 103, 147
Aphanomyces, 197
Apple Valley Sewage Effluent Project, 36, 65
Anthraenose, 102
Aspen, Trembling, 158
Aspergillus spp., 198, 199
Aster sp., 119, 124, 128, 130
Atrazine, 166
Australia, 79, 136
Bacteria, 198
  diseases, 197
  wilt, 101, 197, 199
Balsam fir, 156, 160, 163
Barbarea vulgaris, 130
Barley, 198
Barnyardgrass, 140, 141, 143, 146, 148, 152
Beech, 158-160, 181, 190
Beech-maple, 158-160
Beans, 197
  dry, 15
  navy, 198
Biochemical oxygen demand, 5, 19, 29, 82, 143
Biomass
  accumulation, 126-129
  leaf, 171, 173
  litter, 119, 125, 183, 190, 191
Birdsfoot trefoil, 52, 62, 69, 71, 81, 84-86, 89
Black cherry, 166, 168-169, 171-177, 182
Black walnut, 166, 168-169, 171-177
Boron, 15, 83, 94, 148, 159, 173, 175
Botrytis cinerea, 198
Broadleaf weeds, 89
Bromegrass (smooth), 37, 41, 69, 73, 74, 81, 84-86, 88
Bromus inermis Lyess., 37, 41, 69, 73, 74, 81, 84-86, 88
Brown blotch, 199
Buffer zones, 6, 17
Cadmium, 82
Calcium, 50, 80, 83, 94, 148, 158, 173, 175
Canada, 198
Canada thistle, 130, 191
Candida parapsilosis, 199
Cation exchange capacity (CEC), 6, 136

Center pivot system, 6, 16, 25, 68
Cephalosporium spp., 198
Chaetomium spp., 198
Chloride, 37, 54, 56, 82, 109, 115, 116, 181, 184, 186, 189
Chlorine, 22, 199
Christmas trees, 16, 17, 159, 161-163
Cirsium arvense, 130, 191
Cladosporium cladosporioides, 198
Climate, 23, 27, 76
Clover, 10, 79
  Ladino, 102
  red, 52, 61, 71, 72, 102
Cold Regions Research and Engineering Laboratory, Hanover, NH, 5, 7, 12, 30, 46, 135-137, 145
Coleosporium asterum, 196, 197
Coliform bacteria, 19, 30
Coniothyrium fuckelii, 198
Copper, 83, 94, 148
Corn, 7-10, 12, 14, 24, 28, 36-38, 40-42, 44-46, 49-52, 56-61, 63, 66-73, 76, 81, 83, 85-87, 89, 97-99, 103, 197-198
    ear rot, 197
    fodder, 38, 41
    silage, 59, 70
Coronilla varia, 52, 69
Cotton, 197
Creeping foxtail, 81
Crop
  diseases, 14
  management, 8, 14, 35, 49, 65, 73
  responses, 6, 50, 67
Crown vetch, 52, 69
Curvularia lunata, 198
Dactylis glomerata L., 11, 37, 41, 52, 60, 62, 69, 70, 71, 73-75, 81, 84-86, 88, 89, 91, 94, 97-100, 102, 103, 140, 148
Dairy
  cattle, 101
  manure, 197
Dandelion, 88, 89, 118, 122, 130
Daucus carota, 130
Denitrification, 8, 29, 63, 67, 91, 110, 182, 185
Digitaria, 166
Disinfection, 22, 199
Douglas fir, 156, 157, 162
Dual cropping *(see intercropping)*
Earthworms, 198
Eastern cottonwood, 16, 166, 168, 169, 171-177
Echinochloa crusqualli, 140, 141, 143, 146, 148, 152
Ergot, 197
Erosion, 23, 70
Erwinia herbicola, 200
Erysipe cichoracearum, 196, 198
European larch, 156, 157
Evapotranspiration, 13, 21, 23, 63, 109, 114, 182, 184, 188
Fagus grandifolia, 158-160, 181, 190
Federal Water Pollution Control Act
    Amendments of 1972, 1, 2
    Amendments of 1977, 2, 5
Fertilizer, 67, 82, 95, 137, 140, 162
Festuca arundinacea Schreb., 11, 37, 41, 52, 54, 60, 62, 69, 70, 73-75, 81, 83-86, 89, 99, 102, 137, 140, 148
Food irrigation, 6, 24
Flood processing wastes, 31, 155
Forage
  crops, 65, 80
  grasses, 7, 13, 14, 136, 147
  perennial, 36, 72, 74, 75
  sorghum, 81, 83, 85-87, 89, 91, 94, 96, 103
Forest
  ecosystem, 16, 17
  irrigation, 18, 165, 181, 191

research needs, 18
Fraxinus americana, 157, 166, 168, 169, 171-177
Fraxinus pennsylvanica, 156, 157
Fungus
Alternaria spp., 198
Alternaria tenius, 198
Aspergillus spp., 198, 199
Botrytis cinerea, 198
Cephalosporium spp., 198
Chaetomium spp., 198
Cladosporium cladosporioides, 198
Coleosporium asterum, 196, 197
Coniothyrium fuckelii, 198
Curvularia lunata, 198
Erysipe cichoracearum, 196, 198
Fusarium spp., 197, 199, 198
Gliocladium spp., 198
Helminthosporium sp., 197
Mucor spp., 198, 199
Penicillium spp., 198, 199
Phyllachora graminis, 197
Rhizopus nigricans, 198, 199
Scopulariopsis brevicaulis, 198
Stemphyllium consortiale, 198
Trichoderma viride, 198
Fusarium spp., 197, 199, 198
Georgia, 15
Gibberella, 197
Gliocladium spp., 198
Glycine max Nerr., 9, 15, 198
Goldenrod, 108, 118-120, 128, 130, 166, 196, 197
Grasses, 10, 12, 13, 22, 30, 37, 38, 44-46, 69, 73, 79, 80, 118, 142, 197
annual, 79, 80, 97
forage, 7, 13, 14, 136, 147
perennial, 9, 80, 89, 90, 95, 97, 98, 148
Green ash, 156, 157
Greenville, ME, 16
Great Lakes, 19, 24, 99, 108, 162, 163
Groundwater
aquifer system, 36
contamination, 16, 91, 98
hydrology, 21, 22
quality
monitoring, 53, 56, 58
surveillance, 60, 68
recharge, 7, 184, 188
Hanover, NH, 12, 30
Hardwoods, 16
Harvesting, 13, 30, 102, 110, 111, 114, 120-124, 126, 127, 129, 139, 146, 147, 196
Hay, 22, 63, 72, 101, 146
Heavy metals, 15, 50, 82, 110
Helianthus annuss L., 8
Helminthosporium spp., 197
Herbicides, 9, 19-72, 80, 161, 163, 166
Houghton Lake, MI, 29
Hydrogeologic aspect, 15, 16, 91
Ice, 144
Impatiens capensis, 191
Infiltration, 6,23, 24, 28, 30, 37, 69, 70
rates, 20, 22, 108
Insects, 162
Italian ryegrass, 136
Intercropping, 6, 9, 51, 53, 57, 58, 60, 62, 63, 72, 76
Iron, 82, 83, 94
Irrigation systems *(also see center pivot irrigation; ridge and furrow irrigation; Muskegon County, Michigan; wastewater irrigation project, Michigan State University Water Quality Management Facility; Apple Valley sewage effluent project; Pennsylvania State University Project; and spray irrigation)*
application rates, 14, 16, 17, 23, 25
design, 20
frequency, 15, 16, 29

underdrain, 22
Japanese larch, 156, 157
Juglans nigra L., 166, 168, 169, 171-177
Kentucky bluegrass, 11, 31, 37, 41, 73-76, 81, 84-86, 88, 89, 102, 118, 123, 128, 130, 140, 143, 151
Lactuca canadensis, 130
Lagoon storage, 17, 23, 27, 79, 185
Land treatment design criteria, 6
Larix decidua, 156, 157
Larix leptolepis, 156, 157
Leaching, 7, 9, 12, 13, 44, 66, 69-72, 75, 76, 91, 109, 110, 114, 156, 183, 190
Leaf diseases, 103, 197, 199
Legumes, 7, 9, 24, 60, 79, 70, 73, 74, 79, 80, 82, 95, 96, 97
Lettuce, 197
Linaria vulgaris, 130
Liriodendron tulipifera, 156, 157, 166, 168, 169, 171-177
Litter biomass, 119, 125, 183, 190, 191
Livestock, 65, 80, 97, 99, 100-102
Lolium perenne, 52, 54, 56, 57, 59, 60, 62, 63, 71, 74, 137, 140, 143, 148
Lombardy poplar, 157
Lotus comiculatus, 52, 62, 69, 71, 81, 84-86, 89
Lysimeter, 167
Magnesium, 50, 80, 83, 94, 148, 158, 173, 175
Maine, 16
Manganese, 82, 83, 94, 148, 173, 175
Maple
red, 158-159, 181
sugar, 158-160, 190
Medicago sativa L., 10, 12, 13, 37, 41, 52, 61, 71-75, 81, 83-86, 88-92, 94, 97-99, 101-103, 196, 198, 200
Mercury, 82
Mice, 161
Michigan, 5, 9, 15, 22, 29, 74, 79, 89, 155-157, 159, 160, 165, 181, 183
Michigan State University Water Quality Management Facility, 5, 7, 65, 66-70, 73, 74, 76, 79, 80, 98, 107, 108, 110, 132, 166, 167, 181, 183, 185, 196
Middleville, MI, 26-28
Millet, 197
Milton, WI, 16
Minnesota, 5, 9, 35, 68, 71, 73-75
Mississippi, 136
Mucor spp., 198, 199
Mulching, 11
Municipal wastewater treatment systems, 117, 167, 198
Mushroom, 199
Muskegon, MI, 5, 49-51, 54, 62, 71
Muskegon County, Michigan, wastewater irrigation system, 7, 18, 49-51, 54, 65, 70
Mycorrhizae, 158
National Oceanic and Atmospheric Administration (NOAA), 21
Needle tip necrosis, 159
Nematodes, 197
New Hampshire, 5, 15, 16, 30, 135, 136, 145, 151
Nickel, 82
Nitrates, 7, 12, 16, 49, 54, 57, 63, 68, 70, 75, 80, 82, 91, 99, 100, 107, 109, 112, 115, 116, 156, 161, 167, 177, 183
Nitrification, 29
Nitrogen, 19, 28, 37, 41, 45, 58, 60, 79, 94, 99, 109, 111, 115, 116, 125, 137, 139, 148, 173, 174, 181, 184, 186-188
loading, 51, 138, 146
removal, 7, 12, 13, 17, 18, 41, 49, 50, 62, 63,

66, 68, 70, 73, 75, 89, 95, 96, 97, 149, 150, 152
stripping, 65
Non-row crops, 10
Northern red oak, 156
Northern white cedar, 156, 157
Norway spruce, 157, 166, 168, 169, 171-177
Nutrients
composition, 54
loading rates, 54
removal of crop harvest, 17, 42, 52, 59, 61, 74, 75, 84, 92, 93, 95, 112, 114, 145, 146, 149-151, 170
Oak, 158, 166, 168, 169, 171-177
Oats, 54, 197, 198
Old field, 13, 18, 107, 110, 112, 114, 115, 125, 165
Orchardgrass, 11, 37, 41, 52, 60, 62, 69-71, 73-75, 81, 84-86, 88, 89, 91, 94, 97-100, 102, 103, 140, 148
pennlate, 137
Organic bioaccumulation, 14
Organic matter, 6
Organic nitrogen, 183-185, 188
Organic toxicants, 15, 110
Ornamentals, 16
Overland flow, 5, 10, 14, 16, 19, 20, 23, 28, 29, 31, 135, 136, 141, 142, 149
Oxidation pond, 155, 158
Pack Forest, WA, 16
Paraquat, 57, 80
Parasites, 198
Peas, 197
Penicillium spp., 198, 199
Pennsylvania, 15, 79, 156
Pennsylvania State University, 5, 7, 15, 16, 18, 65, 66, 68-70, 72, 74, 157, 158
Percolation, 13, 24
Phalaris arundinacea L., 11, 12, 31, 37, 40-42, 44, 46, 52, 62, 69-75, 84, 86, 88-90, 97, 99, 102, 130, 136, 137, 140, 143, 148, 196
Phleum pratense Leyss, 37, 41, 69, 73, 74, 81, 84-86, 88, 130
Phloeum pratense L., 137
Phosphate, 37, 100
Phosphorus, 19, 29, 30, 37, 38, 50, 54, 79, 82, 83, 89, 96, 98-100, 107, 109, 114-116, 137, 139, 148-150, 152, 158, 167, 173, 174, 181, 183, 184, 189
loading, 138
removal, 12, 29, 44, 178, 185
Phyllachora graminis, 197
Phytopathogenic bacterium, 200
Phytophthora megasperma, 81, 88, 91, 98, 101
Phytophthora parasitica, 198, 200
Picea abres, 157, 166, 168, 169, 171-177
Picea glauca, 156-158, 160, 163, 166, 168, 169, 171-177
Pine
Austrian, 15, 197
loblolly, 157, 170
pitch, 157
ponderosa, 170
red, 15, 17, 155, 157
scotch, 15, 156, 160, 162, 163, 166, 168, 169, 171-177
Virginia, 158
white, 157, 170
pine-oak, 158
Pinus nigra, 157
Pinus resinosa, 15, 17, 155, 157
Pinus rigida, 157
Pinus strobus, 157, 170
Pinus sylvestris, 16, 156, 160, 162, 163, 166, 168, 169, 171-177
Pinus virginiana L., 158
Plantago spp., 130

Plants
  chlorotic tissue, 196
  diseases, 195
  necrotic tissue, 196
Poa pratensis Leyss, 11, 31, 37, 41, 73, 76, 81, 84-86, 88, 89, 102, 118, 123, 128, 130, 140, 143, 151
Pond storage, 17, 23, 27, 79, 185
Populus, 15, 163
Populus x *euramericana*, 156, 157, 162
Populus canescens x *p. grandidentata*, 156, 157
Populus canescens x *p. tremuloides*, 156, 157
Populus deltoides Bortr., 16, 166, 168, 169, 171-177
Populus nigra var *italica*, 157
Populus tremuloides, 158
Portable solid set, 6, 16, 26, 28
Potassium, 38, 50, 79, 80, 82, 83, 94, 137, 143, 148, 152, 167, 173, 174
Potentilla recta, 130
Preapplication treatment, 19, 30
Protozoans, 198
Prunus serotina Ehrh., 166, 168, 169, 171-177, 182
Pseudomonas tolaasi, 199
Pseudotsuga menziesii, 156, 157, 162
Pythium aphanidermatum, 197, 200
Quackgrass, 13, 37, 41, 52, 62, 71, 73, 88, 89, 107, 108, 118, 121, 128, 136, 140, 143, 151, 166, 196-198
Quercus alba L., 158
Quercus prinus L., 158
Quercus velutina L., 158
Quercus rubra L., 158, 166, 168, 169, 171-177
Quercus coccinea Muenchh., 158
Rabbits, 161
Raccoons, 83
Recharge, 3, 184, 188
Reed canarygrass, 11, 12, 31, 37, 40-42, 44, 46, 52, 62, 69-75, 84, 86, 88-90, 97, 99, 102, 130, 136, 137, 140, 143, 148, 196
Reed foxtail, 74, 84-86
Rhizobium meliloti, 95
Rhizoctonia (damping off), 197
Rhizopus nigricans, 198, 199
Ringspot, 197
Root
  biomass, 7, 9, 67
  depth, 11
  disease, 12
  growth, 22
  rot, 73, 81, 88, 91, 98, 99, 196, 199, 200
  structure, 14, 66, 161, 177
  zone, 12, 24, 91
Row crops, 8, 14
Rumex acetocella, 130
Rumex crispus, 130
Runoff, 8, 23, 70, 114, 188, 190
Rye, 8, 9, 52, 56, 59, 62, 63, 70, 71, 80, 197
Ryegrass
  Italian, 136
  perennial, 52, 54, 56, 57, 59, 60, 62, 63, 71, 74, 137, 140, 143, 148
Sclerotinia (drop), 197
Scopulariopsis brevicaulis, 198
Secale cereale L., 8, 9, 52, 56, 59, 62, 63, 70, 71, 80, 197
Septoria spp., 197
Setaria spp., 130
Shoot:root ratio, 160
Silage, 8, 59, 66, 68, 70, 71, 101
Silviculture, 16, 98
Simazine, 161, 166
Slopes, 26, 29, 32, 138, 142, 151
Sludge, 51, 75, 197, 198
  composting, 17

Sodium, 50, 83, 94, 173, 174
Soil
  aeration, 14
  clogging, 22
  erosion, 8, 140, 151
  nutrients, 15, 22
  permeability, 22, 24, 100
  pH, 136
  physical and chemical characteristics of, 6, 20, 22, 29, 31, 51, 80, 83, 96, 96, 107, 136, 158, 160, 166, 167, 182,
  removal efficiency, 30
  renovation processes, 111
  sampling, 53
  structure, 22, 135
  texture, 20, 23, 135
  temperature, 32
  water, 37, 44, 66, 75, 76, 83, 91, 95, 96, 98-101, 187
Soil Conservation Service, 20, 21, 24
Solidago graminifoli; canadensis, 108, 118-120, 128, 130, 166, 196, 197
Sorghum
  forage, 81, 83, 85-87, 89, 91, 94, 96, 103
  sudangrass, 81, 83, 85-87, 89, 103
Soybeans, 9, 15, 198
Spray irrigation, 5, 6, 107, 109, 181, 182
  facilities, 9
  systems, 10, 12
Sprinklers, 159, 163
Stabilization ponds, 19
Stagonospora foliicola, 196
Stemphyllium consortiali, 198
Stewart's bacterial wilt, 197
Storage requirements, 30
Suspended solids, 19, 82
Sunflower, 8
Tall fescue, 11, 37, 41, 52, 54, 60, 62, 69-71, 73-75, 81, 83-86, 88, 89, 99, 102, 137, 140, 148
Taraxacum officinale L., 88, 89, 118, 122, 130
Tawny blotch, 196
Texas, 136
Thielavioposis, 197
Thuja occidentalis, 156, 157
Tilia americana, 181
Timothy, 37, 41, 69, 73, 74, 81, 84-86, 88, 130
Trace metal toxicity, 147
Traveler gun irrigation system, 25
Trees, 71
  diseases, 162
  growth response, 17, 155, 156, 165, 168, 170, 176, 183, 190
  mortality, 171
  seedlings, 15
  wood chips, 16, 162
Trembling aspen, 158
Trichoderma viride, 198
Trifolium spp., 10, 79
Trifolium pratense, 52, 61, 71, 72, 102
Trifolium repens, 102
Ulmus americana, 181
Ulmus rubra, 181, 190
United States Army Corps of Engineers, 2, 5
United States Department of Agriculture, 2, 5
United States Department of the Interior, 132
United States Environmental Protection Agency, 1-3, 21, 29, 98, 99, 132
United States Forest Service, 5
United States Geological Service, 21
Vegetation management, 14
Vermont, 16
Vermontville, MI, 29
Viruses, 199
Washington, 15, 16
Wastewater
  application rate, 67, 68, 73, 75, 82, 89, 90, 95-97, 99-101, 109, 111, 118, 120-124, 126, 127, 129,

184, 187, 191
characteristics, 39, 55, 82
construction grants, 2
financial incentives, 2
food processing, 30, 155
industrial, 58
inovative technology, 1, 5
tertiary treatment, 107
Water quality, 41
Water renovated, 177
Weeds, broad leafed, 89
West Dover, VT, 16
Wetlands, 28, 29
Wheat, 16, 157, 166, 168, 169, 171-177
White spruce, 156-158, 160, 163, 166, 168, 169, 171-177
Winter, 143, 144
cover, 69, 102
irrigation, 7, 16-18, 23, 73, 107-109
Wisconsin, 16
Wolfboro, NH 16
Yeast, 199, 200
Zea mays L., 7-10, 12, 14, 24, 28, 36-38, 40-42, 44-46, 49-52, 56-61, 63, 66-73, 76, 81, 85-87, 89, 97-99, 103, 197, 198
Zinc, 82, 83, 94, 148, 173, 175